Astronomy for Amateurs

by Camille Flammarion

INTRODUCTION

The Science of Astronomy is sublime and beautiful. Noble, elevating, consoling, divine, it gives us wings, and bears us through Infinitude. In these ethereal regions all is pure, luminous, and splendid. Dreams of the Ideal, even of the Inaccessible, weave their subtle spells upon us. The imagination soars aloft, and aspires to the sources of Eternal Beauty.

What greater delight can be conceived, on a fine spring evening, at the hour when the crescent moon is shining in the West amid the last glimmer of twilight, than the contemplation of that grand and silent spectacle of the stars stepping forth in sequence in the vast Heavens? All sounds of life die out upon the earth, the last notes of the sleepy birds have sunk away, the Angelus of the church hard by has rung the close of day. But if life is arrested around us, we may seek it in the Heavens. These incandescing orbs are so many points of interrogation suspended above our heads in the inaccessible depths of space.... Gradually they multiply. There is Venus, the white star of the shepherd. There Mars, the little celestial world so near our own. There the giant Jupiter. The seven stars of the Great Bear seem to point out the pole, while they slowly revolve around it.... What is this nebulous light that blanches the darkness of the heavens, and traverses the constellations like a celestial path? It is the Galaxy, the Milky Way, composed of millions on millions of suns!... The darkness is profound, the abyss immense.... See! Yonder a shooting star glides silently across the sky, and disappears!...

Who can remain insensible to this magic spectacle of the starry Heavens? Where is the mind that is not attracted to these enigmas? The intelligence of the amateur, the feminine, no less than the more material and prosaic masculine mind, is well adapted to the consideration of astronomical problems. Women, indeed, are naturally predisposed to these contemplative studies. And the part they are called to play in the education of our children is so vast, and so important, that the elements of Astronomy might well be taught by the young mother herself to the budding minds that are curious about every issue-- whose first impressions are so keen and so enduring.

Throughout the ages women have occupied themselves successfully with Astronomy, not merely in its contemplative and descriptive, but also in its mathematical aspects. Of such, the most illustrious was the beautiful and learned Hypatia of Alexandria, born in the year 375 of our era, public lecturer on geometry, algebra, and astronomy, and author of three works of great importance. Then, in that age of ignorance and fanaticism, she fell a victim to human stupidity and malice, was dragged from her chariot while crossing the Cathedral Square, in March, 415, stripped of her garments, stoned to death, and burned as a dishonored witch!

Among the women inspired with a passion for the Heavens may be cited St. Catherine of Alexandria, admired for her learning, her beauty and her virtue. She was martyred in the reign of Maximinus Daza, about the year 312, and has given her name to one of the lunar rings.

Another celebrated female mathematician was Madame Hortense Lepaute, born in 1723, who collaborated with Clairaut in the immense calculations by which he predicted the return of Halley's Comet. "Madame Lepaute," wrote Lalande, "gave us such immense assistance that, without her, we should never have ventured to undertake this enormous labor, in which it was necessary to calculate for every degree, and for a hundred and fifty years, the distances and forces of the planets acting by their attraction on the comet. During more than six months, we calculated from morning to night, sometimes even at table, and as the result of this forced labor I contracted an illness that has changed my constitution for life; but it was important to publish the result before the arrival of the comet."

This extract will suffice for the appreciation of the scientific ardor of Madame Lepaute. We are indebted to her for some considerable works. Her husband was clock-maker to the King. "To her intellectual talents," says one of her biographers, "were joined all the qualities of the heart. She was charming to a degree, with an elegant figure, a dainty foot, and such a beautiful hand that Voiriot, the King's painter, who had made a portrait of her, asked permission to copy it, in order to preserve a model of the best in Nature." And then we are told that learned women can not be good-looking!...

The Marquise du Chelet was no less renowned. She was predestined to her career, if the following anecdote be credible. Gabrielle de Breteuil, born in 1706 (who, in 1725, was to marry the Marquis du Chelet, becoming, in 1733, the most celebrated friend of Voltaire), was four or five years old when she was given an old compass, dressed up as a doll, for a plaything. After examining this object for some time, the child began angrily and impatiently to strip off the silly draperies the toy was wrapped in, and after turning it over several times in her little hands, she divined its uses, and traced a circle with it on a sheet of paper. To her, among other things, we owe a precious, and indeed the only French, translation of Newton's great work on universal gravitation, the famous Principia, and she was, with Voltaire, an eloquent propagator of the theory of attraction, rejected at that time by the Acadamie des Sciences.

Numbers of other women astronomers might be cited, all showing how accessible this highly abstract science is to the feminine intellect. President des Brosses, in his charming Voyage en Italie, tells of the visit he paid in Milan to the young Italian, Marie Agnesi, who delivered harangues in Latin, and was acquainted with seven languages, and for whom mathematics held no secrets. She was devoted to algebra and geometry, which, she said, "are the only provinces of thought wherein peace reigns." Madame de Charrie expressed herself in an aphorism of the same order: "An hour or two of mathematics sets my mind at liberty, and puts me in good spirits; I feel that I can eat and sleep better when I have seen obvious and indisputable truths. This consoles me for the obscurities of religion and metaphysics, or rather makes me forget them; I am thankful there is something positive in this world." And did not Madame de Blocqueville, last surviving daughter of Marshal Davout, who died in 1892, exclaim in her turn: "Astronomy, science of sciences! by which I am attracted, and terrified, and which I adore! By it my soul is detached from the things of this world, for it draws me to those unknown spheres that evoked from Newton the triumphant cry: 'Coeli enarrant gloriam Dei!'"

Nor must we omit Miss Caroline Herschel, sister of the greatest observer of the Heavens, the grandest discoverer of the stars, that has ever lived. Astronomy gave her a long career; she discovered no less than seven comets herself, and her patient labors preserved her to the age of ninety-eight.--And Mrs. Somerville, to whom we owe the English translation of Laplace's Celeste, of whom Humboldt said, "In pure mathematics, Mrs. Somerville is absolutely superior." Like Caroline Herschel, she was almost a centenarian, appearing always much younger than her years: she died at Naples, in 1872, at the age of ninety-two.--So, too, the Russian Sophie Kovalevsky, descendant of Mathias Corvinus, King of Hungary, who, an accomplished mathematician at sixteen, married at eighteen, in order to follow the curriculum at the University (then forbidden to unmarried women); arranging with her young husband to live as brother and sister until their studies should be completed. In 1888 the Prix Bordin of the Institut was conferred on her.--And Maria Mitchell of the United States, for whom Le Verrier gave a f 階 e at the Observatory of Paris, and who was exceptionally authorized by Pope Pius IX to visit the Observatory of the Roman College, at that time an ecclesiastical establishment, closed to women.--And Madame Scarpellini, the Roman astronomer, renowned for her works on shooting stars, whom the author had the honor of visiting, in company with Father Secchi, Director of the Observatory mentioned above.

At the present time, Astronomy is proud to reckon among its most famous workers Miss Agnes Clerke, the learned Irishwoman, to whom we owe, inter alia, an excellent History of Astronomy in the Nineteenth Century;--Mrs. Isaac Roberts, who, under the familiar name of Miss Klumpke, sat on the Council of the Astronomical Society of France, and is D. Sc. of the Faculty of Paris and head of the Bureau for measuring star photographs at the Observatory of Paris (an American who became English by her marriage with the astronomer Roberts, but is not forgotten in France);--Mrs. Fleming, one of the astronomers of the Observatory at Harvard College, U.S.A., to whom we owe the discovery of a great number of variable stars by the examination of photographic records, and by spectral photography;--Lady Huggins, who in England is the learned collaborator of her illustrious husband;--and many others.

* * * * *

The following chapters, which aim at summing up the essentials of Astronomy in twelve lessons for amateurs, will not make astronomers or mathematicians of my readers--much less prigs or pedants. They are designed to show the constitution of the Universe, in its grandeur and its beauty, so that, inhabiting this world, we may know where we are living, may realize our position in the Cosmos, appreciate Creation as it is, and enjoy it to better advantage. This sun by which we live, this succession of months and years, of days and nights, the apparent motions of the heavens, these starry skies, the divine rays of the moon, the whole totality of things, constitutes in some sort the tissue of our existence, and it is indeed extraordinary that the inhabitants of our planet should almost all have lived till now without knowing where they are, without suspecting the marvels of the Universe.

* * * * *

For the rest, my little book is dedicated to a woman, muse and goddess--the charming enchantress Urania, fit companion of Venus, ranking even above her in the choir of celestial beauties, as purer and more noble, dominating with her clear glance the immensities of the universe. Urania, be it noted, is feminine, and never would the poetry of the ancients have imagined a masculine symbol to personify the pageant of the heavens. Not Uranus, nor Saturn, nor Jupiter can compare with the ideal beauty of Urania.

Moreover, I have before me two delightful books, in breviary binding, dated the one from the year 1686, the other from a century later, 1786. The first was written by Fontenelle for a Marquise, and is entitled Entretiens sur la Pluralit?des Mondes. In this, banter is pleasantly married with science, the author declaring that he only demands from his fair readers the amount of application they would concede to a novel. The second is written by Lalande, and is called Astronomie des Dames. In addressing myself to both sexes, I am in honorable company with these two sponsors and esteem myself the better for it.

CHAPTER I

THE CONTEMPLATION OF THE HEAVENS

The crimson disk of the Sun has plunged beneath the Ocean. The sea has decked itself with the burning colors of the orb, reflected from the Heavens in a mirror of turquoise and emerald. The rolling waves are gold and silver, and break noisily on a shore already darkened by the disappearance of the celestial luminary.

We gaze regretfully after the star of day, that poured its cheerful rays anon so generously over many who were intoxicated with gaiety and happiness. We dream, contemplating the magnificent spectacle, and in dreaming forget the moments that are rapidly flying by. Yet the darkness gradually increases, and twilight gives way to night.

The most indifferent spectator of the setting Sun as it descends beneath the waves at the far horizon, could hardly be unmoved by the pageant of Nature at such an impressive moment.

The light of the Crescent Moon, like some fairy boat suspended in the sky, is bright enough to cast changing and dancing sparkles of silver upon the ocean. The Evening Star declines slowly in its turn toward the western horizon. Our gaze is held by a shining world that dominates the whole of the occidental heavens. This is the "Shepherd's Star," Venus of rays translucent.

Little by little, one by one, the more brilliant stars shine out. Here are the white Vega of the Lyre, the burning Arcturus, the seven stars of the Great Bear, a whole sidereal population catching fire, like innumerable eyes that open on the Infinite. It is a new life that is revealed to our imagination, inviting us to soar into these mysterious regions.

O Night, diapered with fires innumerable! hast thou not written in flaming letters on these Constellations the syllables of the great enigma of Eternity? The contemplation of thee is a wonder and a charm. How rapidly canst thou efface the regrets we suffered on the departure of our beloved Sun! What wealth, what beauty hast thou not reserved for our enraptured souls! Where is the man that can remain blind to such a pageant and deaf to its language!

To whatever quarter of the Heavens we look, the splendors of the night are revealed to our astonished gaze. These celestial eyes seem in their turn to gaze at, and to question us. Thus indeed have they questioned every thinking soul, so long as Humanity has existed on our Earth. Homer saw and sung these self-same stars. They shone upon the slow succession of civilizations that have disappeared, from Egypt of the period of the Pyramids, Greece at the time of the Trojan War, Rome and Carthage, Constantine and Charlemagne, down to the Twentieth Century. The generations are buried with the dust of their ancient temples. The Stars are still there, symbols of Eternity.

The silence of the vast and starry Heavens may terrify us; its immensity may seem to overwhelm us. But our inquiring thought flies curiously on the wings of dream, toward the remotest regions of the visible. It rests on one star and another, like the butterfly on the flower. It seeks what will best respond to its aspirations: and thus a kind of communication is established, and, as it were, protected by all Nature in these silent appeals. Our sense of solitude has disappeared. We feel that, if only as infinitesimal atoms, we form part of that immense universe, and this dumb language of the starry night is more eloquent than any speech. Each star becomes a friend, a discreet confidant, often indeed a precious counsellor, for all the thoughts it suggests to us are pure and holy.

Is any poem finer than the book written in letters of fire upon the tablets of the firmament? Nothing could be more ideal. And yet, the poetic sentiment that the beauty of Heaven awakens in our soul ought not to veil its reality from us. That is no less marvelous than the mystery by which we were enchanted.

And here we may ask ourselves how many there are, even among thinking human beings, who ever raise their eyes to the starry heavens? How many men and women are sincerely, and with unfeigned curiosity, interested in these shining specks, and inaccessible luminaries, and really desirous of a better acquaintance with them?

Seek, talk, ask in the intercourse of daily life. You, who read these pages, who already love the Heavens, and comprehend them, who desire to account for our existence in this world, who seek to know what the Earth is, and what Heaven--you shall witness that the number of those inquiring after truth is so limited that no one dares to speak of it, so disgraceful is it to the so-called intelligence of our race. And yet! the great Book of the Heavens is open to all eyes. What pleasures await us in the study of the Universe! Nothing could speak more eloquently to our heart and intellect!

Astronomy is the science par excellence. It is the most beautiful and most ancient of all, inasmuch as it dates back to the indeterminate times of highest antiquity. Its mission is not only to make us acquainted with the innumerable orbs by which our nights are illuminated, but it is, moreover, thanks to it that we know where and what we are. Without it we should live as the blind, in eternal ignorance of the very conditions of our terrestrial existence. Without it we should still be penetrated with the naive error that reduced the entire Universe to our minute globule, making our Humanity the goal of the Creation, and should have no exact notion of the immense reality.

To-day, thanks to the intellectual labor of so many centuries, thanks also to the immortal genius of the men of science who have devoted their lives to searching after Truth--men such as Copernicus, Galileo, Kepler, Newton--the veil of ignorance has been rent, and glimpses of the marvels of creation are perceptible in their splendid truth to the dazzled eye of the thinker.

The study of Astronomy is not, as many suppose, the sacrifice of oneself in a cerebral torture that obliterates all the beauty, the fascination, and the grandeur of the pageant of Nature. Figures, and naught but figures, would not be entertaining, even to those most desirous of instruction. Let the reader take courage! We do not propose that he shall decipher the hieroglyphics of algebra and geometry. Perish the thought! For the rest, figures are but the scaffolding, the method, and do not exist in Nature.

We simply beg of you to open your eyes, to see where you are, so that you may not stray from the path of truth, which is also the path of happiness. Once you have entered upon it, no persuasion will be needed to make you persevere. And you will have the profound satisfaction of knowing that you are thinking correctly, and that it is infinitely better to be educated than to be ignorant. The reality is far beyond all dreams, beyond the most fantastic imagination. The most fairy-like transformations of our theaters, the most resplendent pageants of our military reviews, the most sumptuous marvels on which the human race can pride itself--all that we admire, all that we envy on the Earth--is as nothing compared with the unheard-of wonders scattered through Infinitude. There are so many that one does not know how to see them. The fascinated eye would fain grasp all at once.

If you will yield yourselves to the pleasure of gazing upon the sparkling fires of Space, you will never regret the moments passed all too rapidly in the contemplation of the Heavens.

Diamonds, turquoises, rubies, emeralds, all the precious stones with which women love to deck themselves, are to be found in greater perfection, more beautiful, and more splendid, set in the immensity of Heaven! In the telescopic field, we may watch the progress of armies of majestic and powerful suns, from whose attacks there is naught to fear. And these vagabond comets and shooting stars and stellar nebul? do they not make up a prodigious panorama? What are our romances in comparison with the History of Nature? Soaring toward the Infinite, we purify our souls from all the baseness of this world, we strive to become better and more intelligent.

* * * * *

But in the first place, you ask, what are the Heavens? This vault oppresses us. We can not venture to investigate it.

Heaven, we reply, is no vault, it is a limitless immensity, inconceivable, unfathomable, that surrounds us on all sides, and in the midst of which our globe is floating. THE HEAVENS ARE ALL THAT EXISTS, all that we see, and all that we do not see: the Earth on which we are, that bears us onward in her rapid flight; the Moon that accompanies us, and sheds her soft beams upon our silent nights; the good Sun to which we owe our existence; the Stars, suns of Infinitude; in a word--the whole of Creation.

Yes, our Earth is an orb of the Heavens: the sky is her domain, and our Sun, shining above our heads, and fertilizing our seasons, is as much a star as the pretty sparkling points that scintillate up there, in the far distance, and embellish the calm of our nights with their brilliancy. All are in the Heavens, you as well as I, for the Earth, in her course through Space, bears us with herself into the depths of Infinitude.

In the Heavens there is neither "above" nor "below." These words do not exist in celestial speech, because their significance is relative to the surface of this planet only. In reality, for the inhabitants of the Earth, "low" is the inside, the center of the globe, and "high" is what is above our heads, all round the Earth. The Heavens are what surround us on all sides, to Infinity.

The Earth is, like her fellows, Mercury, Venus, Mars, Jupiter, Saturn, Uranus, Neptune, one of the planets of the great solar family.

The Sun, her father, protects her, and directs all her actions. She, as the grateful daughter, obeys him blindly. All float in perfect harmony over the celestial ocean.

But, you may say, on what does the Earth rest in her ethereal navigation?

On nothing. The Earth turns round the colossal Sun, a little globe of relatively light weight, isolated on all sides in Space, like a soap-bubble blown by some careless child.

Above, below, on all sides, millions of similar globes are grouped into families, and form other systems of worlds revolving round the numerous and distant stars that people Infinitude; suns more or less analogous to that by which we are illuminated, and generally speaking of larger bulk, although our Sun is a million times larger than our planet.

Among the ancients, before the isolation of our globe in Space and the motions that incessantly alter its position were recognized, the Earth was supposed to be the immobile lower half of the Universe. The sky was regarded as the upper half. The ancients supplied our world with fantastic supports that penetrated to the Infernal Regions. They could not admit the notion of the Earth's isolation, because they had a false idea of its weight. To-day, however, we know positively that the Earth is based on nothing. The innumerable journeys accomplished round it in all directions give definite proof of this. It is attached to nothing. As we said before, there is neither "above" nor "below" in the Universe. What we call "below" is the center of the Earth. For the rest the Earth turns upon its own axis in twenty-four hours. Night is only a partial phenomenon, due to the rotary motion of the planet, a motion that could not exist under conditions other than that of the absolute isolation of our globe in space.

Since the Sun can only illuminate one side of our globe at one moment, that is to say one hemisphere, it follows that Night is nothing but the state of the part that is not illuminated. As the Earth revolves upon itself, all the parts successively exposed to the Sun are in the day, while the parts situated opposite to the Sun, in the cone of shadow produced by the Earth itself, are in night. But whether it be noon or midnight, the stars always occupy the same position in the Heavens, even when, dazzled by the ardent light of the orb of day, we can no longer see them; and when we are plunged into the darkness of the night, the god Phoebus still continues to pour his beneficent rays upon the countries turned toward him.

The sequence of day and night is a phenomenon belonging, properly speaking, to the Earth, in which the rest of the Universe does not participate. The same occurs for every world that is illuminated by a sun, and endowed with a rotary movement. In absolute space, there is no succession of nights and days.

Upheld in space by forces that will be explained at a later point, our planet glides in the open heavens round our Sun.

Imagine a magnificent aerostat, lightly and rapidly cleaving space. Surround it with eight little balloons of different sizes, the smallest like those sold on the streets for children to play with, the larger, such as are distributed for a bonus in large stores. Imagine this group sailing through the air, and you have the system of our worlds in miniature.

Still, this is only an image, a comparison. The balloons are held up by the atmosphere, in which they float at equilibrium. The Earth is sustained by nothing material. What maintains her in equilibrium is the ethereal void; an immaterial force; gravitation. The Sun attracts her, and if she did not revolve, she would drop into him; but rotating round him, at a speed of 107,000 kilometers[2] (about 66,000 miles) per hour, she produces a centrifugal force, like that of a stone in a sling, that is precisely equivalent, and of contrary sign, to its gravitation toward the central orb, and these two equilibrated forces keep her at the same medium distance.

This solar and planetary group does not exist solitary in the immense void that extends indefinitely around us. As we said above, each star that we admire in the depths of the sky, and to which we lift up our eyes and thoughts during the charmed hours of the night, is another sun burning with its own light, the chief of a more or less numerous family, such as are multiplied through all space to infinity. Notwithstanding the immense distances between the sun-stars, Space is so vast, and the number of these so great, that by an effect of perspective due solely to the distance, appearances would lead us to believe that the stars were touching. And under certain telescopic aspects, and in some of the astral photographs, they really do appear to be contiguous.

The Universe is infinite. Space is limitless. If our love for the Heavens should incite in us the impulse, and provide us with the means of undertaking a journey directed to the ends of Heaven as its goal, we should be astonished, on arriving at the confines of the Milky Way, to see the grandiose and phenomenal spectacle of a new Universe unfold before our dazzled eyes; and if in our mad career we crossed this new archipelago of worlds to seek the barriers of Heaven beyond them, we should still find universe eternally succeeding to universe before us. Millions of suns roll on in the immensities of Space. Everywhere, on all sides, Creation renews itself in an infinite variety.

According to all the probabilities, universal life is distributed there as well as here, and has sown the germ of intelligence upon those distant worlds that we divine in the vicinity of the innumerable suns that plow the ether, for everything upon the Earth tends to show that Life is the goal of Nature. Burning foci, inextinguishable sources of warmth and light, these various, multi-colored suns shed their rays upon the worlds that belong to them and which they fertilize.

Our globe is no exception in the Universe. As we have seen, it is one of the celestial orbs, nourished, warmed, lighted, quickened by the Sun, which in its turn again is but a star.

Innumerable Worlds! We dream of them. Who can say that their unknown inhabitants do not think of us in their turn, and that Space may not be traversed by waves of thought, as it is by the vibrations of light and universal gravitation? May not an immense solidarity, hardly guessed at by our imperfect senses, exist between the Celestial Humanities, our Earth being only a modest planet.

Let us meditate on this Infinity! Let us lose no opportunity of employing the best of our hours, those of the silence and peace of the bewitching nights, in contemplating, admiring, spelling out the words of the Great Book of the Heavens. Let our freed souls fly swift and rapt toward those marvelous countries where indescribable joys are prepared for us, and let us do homage to the first and most splendid of the sciences, to Astronomy, which diffuses the light of Truth within us.

To poetical souls, the contemplation of the Heavens carries thought away to higher regions than it attains in any other meditation. Who does not remember the beautiful lines of Victor Hugo in the Orientales? Who has not heard or read them? The poem is called "Ecstasy," and it is a fitting title. The words are sometimes set to music, and the melody seems to complete their pure beauty:

Note: Free Translation

I was alone on the waves, on a starry night, Not a cloud in the sky, not a sail in sight, My eyes pierced beyond the natural world... And the woods, and the hills, and the voice of Nature Seemed to question in a confused murmur, The waves of the Sea, and Heaven's fires.

And the golden stars in infinite legion, Sang loudly, and softly, in glad recognition, Inclining their crowns of fire;... And the waves that naught can check nor arrest Sang, bowing the foam of their haughty crest... Behold the Lord God--Jehovah!

The immortal poet of France was an astronomer. The author more than once had the honor of conversing with him on the problems of the starry sky--and reflected that astronomers might well be poets.

It is indeed difficult to resist a sense of profound emotion before the abysses of infinite Space, when we behold the innumerable multitude of worlds suspended above our heads. We feel in this solitary contemplation of the Heavens that there is more in the Universe than tangible and visible matter: that there are forces, laws, destinies. Our ants' brains may know themselves microscopic, and yet recognize that there is something greater than the Earth, the Heavens;--more absolute than the Visible, the Invisible;--beyond the more or less vulgar affairs of life, the sense of the True, the Good, the Beautiful. We feel that an immense mystery broods over Nature,--over Being, over created things. And it is here again that Astronomy surpasses all the other sciences, that it becomes our sovereign teacher, that it is the pharos of modern philosophy.

O Night, mysterious, sublime, and infinite! withdrawing from our eyes the veil spread above us by the light of day, giving back transparency to the Heavens, showing us the prodigious reality, the shining casket of the celestial diamonds, the innumerable stars that succeed each other interminably in immeasurable space! Without Night we should know nothing. Without it our eyes would never have divined the sidereal population, our intellects would never have pierced the harmony of the Heavens, and we should have remained the blind, deaf parasites of a world isolated from the rest of the universe. O Sacred Night! If on the one hand it rests upon the heights of Truth beyond the day's illusions, on the other its invisible urns pour down a silent and tranquil peace, a penetrating calm, upon our souls that weary of Life's fever. It makes us forget the struggles, perfidies, intrigues, the miseries of the hours of toil and noisy activity, all the conventionalities of civilization. Its domain is that of rest and dreams. We love it for its peace and calm tranquillity. We love it because it is true. We love it because it places us in communication with the other worlds, because it gives us the presage of Life, Universal and Eternal, because it brings us Hope, because it proclaims us citizens of Heaven.

CHAPTER II

THE CONSTELLATIONS

In Chapter I we saw the Earth hanging in space, like a globe isolated on all sides, and surrounded at vast distances by a multitude of stars.

These fiery orbs are suns like that which illuminates ourselves. They shine by their own light. We know this for a fact, because they are so far off that they could neither be illuminated by the Sun, nor, still more, reflect his rays back upon us: and because, on the other hand, we have been able to measure and analyze their light. Many of these distant suns are simple and isolated; others are double, triple, or multiple; others appear to be the centers of systems analogous to that which gravitates round our own Sun, and of which we form part. But these celestial tribes are situated at such remote distances from us that it is impossible to distinguish all the individuals of each particular family. The most delicate observations have only revealed a few of them. We must content ourselves here with admiring the principals,--the sun-stars,--prodigious globes, flaming torches, scattered profusely through the firmament.

How, then, is one to distinguish them? How can they be readily found and named? There are so many of them!

Do not fear; it is quite a simple matter. In studying the surface of the Earth we make use of geographical maps on which the continents and seas of which it consists are drawn with the utmost care. Each country of our planet is subdivided into states, each of which has its proper name. We shall pursue the same plan in regard to the Heavens, and it will be all the easier since the Great Book of the Firmament is constantly open to our gaze. Our globe, moreover, actually revolves upon itself so that we read the whole in due sequence. Given a clear atmosphere, and a little stimulus to the will from our love of truth and science, and the geography of the Heavens, or "uranography," will soon be as familiar to us as the geography of our terrestrial atom.

On a beautiful summer's night, when we look toward the starry sky, we are at first aware only of a number of shining specks. The stars seem to be scattered almost accidentally through Space; they are so numerous and so close to one another that it would appear rash to attempt to name them separately. Yet some of the brighter ones particularly attract and excite our attention. After a little observation we notice a certain regularity in the arrangement of these distant suns, and take pleasure in drawing imaginary figures round the celestial groups.

That is what the ancients did from a practical point of view. In order to guide themselves across the trackless ocean, the earliest Phenician navigators noted certain fixed bearings in the sky, by which they mapped out their routes. In this way they discovered the position of the immovable Pole, and acquired empire over the sea. The Chaldean pastors, too, the nomad people of the East, invoked the Heavens to assist in their migrations. They grouped the more brilliant of the stars into Constellations with simple outlines, and gave to each of these celestial provinces a name derived from mythology, history, or from the natural kingdoms. It is impossible to determine the exact epoch of this primitive celestial geography. The Centaur Chiron, Jason's tutor, was reputed the first to divide the Heavens upon the sphere of the Argonauts. But this origin is a little mythical! In the Bible we have the Prophet Job, who names Orion, the Pleiades, and the Hyades, 3,300 years ago. The Babylonian Tables, and the hieroglyphs of Egypt, witness to an astronomy that had made considerable advance even in those remote epochs. Our actual constellations, which are doubtless of Babylonian origin, appear to have been arranged in their present form by the learned philosopher Eudoxus of Cnidus, about the year 360 B.C. Aratus sang of them in a didactic poem toward 270. Hipparchus of Rhodes was the first to note the astronomical positions with any precision, one hundred and thirty years before our era. He classified the stars in order of magnitude, according to their apparent brightness; and his catalogue, preserved in the Almagest of Ptolemy, contains 1,122 stars distributed into forty-eight Constellations.

The figures of the constellations, taken almost entirely from fable, are visible only to the eyes of the imagination, and where the ancients placed such and such a person or animal, we may see, with a little good-will, anything we choose to fancy. There is nothing real about these figures. And yet it is indispensable to be able to recognize the constellations in order to find our way among the innumerable army of the stars, and we shall commence this study with the description of the most popular and best known of them all, the one that circles every night through our Northern Heavens. Needless to name it; it is familiar to every one. You have already exclaimed--the Great Bear!

This vast and splendid association of suns, which is also known as the Chariot of David, the Plow or Charles's Wain, and the Dipper, is one of the finest constellations in the Heavens, and one of the oldest--seeing that the Chinese hailed it as the divinity of the North, over three thousand years ago.

If any of my readers should happen to forget its position in the sky, the following is a very simple expedient for finding it. Turn to the North--that is, opposite to the point where the sun is to be found at midday. Whatever the season of the year, day of the month, or hour of the night, you will always see, high up in the firmament, seven magnificent stars, arranged in a quadrilateral, followed by a tail, or handle, of three stars. This magnificent constellation never sinks below our horizon. Night and day it watches above us, turning in twenty-four hours round a very famous star that we shall shortly become acquainted with. In the figure of the Great Bear, the four stars of the quadrilateral are found in the body, and the three at the extremity make the tail. As David's Chariot, the four stars represent the wheels, and the three others the horses.

Sometimes our ancestors called them the Seven Oxen, the "oxen of the celestial pastures," from which the word septentrion (septem triones, seven oxen of labor) is derived. Some see a Plowshare; others more familiarly call this figure the Dipper. As it rotates round the pole, its outline varies with the different positions.

It is not easy to guess why this constellation should have been called the Bear. Yet the name has had a certain influence. From the Greek word arctos (bear) has come arctic, and for its antithesis, antarctic. From the Latin word trio (ox of labor) has come septentrion, the seven oxen. Etymology is not always logical. Is not the word "venerate" derived from Venus?

In order to distinguish one star from another, the convention of denoting them by the letters of the Greek Alphabet has been adopted, for it would be impossible to give a name to each, so considerable is their number.[3]

[alpha] and [beta] denote the front wheels of the Chariot generally known as the "pointers;" [gamma] and [delta] the hind wheels; [epsilon], [zeta], [eta] the three horses. All these stars are of the second order of magnitude (the specific meaning of this expression will be explained in the next chapter), except the last ([delta]) of the quadrilateral, which is of the third order.

Figure 3 gives the outline of this primitive constellation. In revolving in twenty-four hours round the Pole, which is situated at the prolongation of a line drawn from [beta] to [alpha], it occupies every conceivable position,--as if this page were turned in all directions. But the relative arrangement of the seven stars remains unaltered. In contemplating these seven stars it must never be forgotten that each is a dazzling sun, a center of force and life. One of them is especially remarkable: [zeta], known as Mizar to the Arabs. Those who have good sight will distinguish near it a minute star, Alcor, or the Cavalier, also called Saidak by the Arabs--that is, the Test, because it can be used as a test of vision. But further, if you have a small telescope at your disposal, direct it upon the fine star Mizar: you will be astonished at discovering two of the finest diamonds you could wish to see, with which no brilliant is comparable.

There are several double stars; these we shall become acquainted with later on.

Meantime, we must not forget our celestial geography. The Great Bear will help us to find all the adjacent constellations.

If a straight line is drawn (Fig. 4) from [beta] through [alpha], which forms the extremity of the square, and is prolonged by a quantity equal to the distance of [alpha] from the tip of the handle, we come on a star of second magnitude, which marks the extremity of a figure perfectly comparable with the Great Bear, but smaller, less brilliant, and pointing in the contrary direction. This is the Little Bear, composed, like its big brother, of seven stars; the one situated at the end of the line by which we have found it is the Pole-Star.

Immovable in the region of the North Pole, the Pole-Star has captivated all eyes by its position in the firmament. It is the providence of mariners who have gone astray on the ocean, for it points them to the North, while it is the pivot of the immense rotation accomplished round it by all the stars in twenty-four hours. Hence it is a very important factor, and we must hasten to find it, and render it due homage. It should be added that its special immobility, in the prolongation of the Earth's axis, is merely an effect caused by the diurnal movements of our planet. Our readers are of course aware that it is the earth that turns and not the sky. But evidence of this will be given later on. In looking at the Pole-Star, the South is behind one, the East to the right, and the West to the left.

Between the Great and the Little Bear, we can distinguish a winding procession of smaller stars. These constitute the Dragon.

We will continue our journey by way of Cassiopeia, a fine constellation placed on the opposite side of the Pole-Star in relation to the Great Bear, and shaped somewhat like the open limbs of the letter W. It is also called the Chair. And, in fact, when the figure is represented with the line [alpha] [beta] below, the line [chi] [gamma] forms the seat, and [gamma] [delta] [epsilon] its back.

If a straight line is drawn from [delta] of the Great Bear, and prolonged beyond the Pole-Star in a quantity equal to the distance which separates these two stars, it is easy to find this constellation (Fig. 5). This group, like the preceding, never sets, and is always visible, opposite to the Great Bear. It revolves in twenty-four hours round the Pole-Star, and is to be seen, now above, now below, now to the right, now to the left.

If in the next place, starting from the stars [alpha] and [delta] in the Great Bear, we draw two lines which join at Polaris and are prolonged beyond Cassiopeia, we arrive at the Square of Pegasus (Fig. 6), a vast constellation that terminates on one side in a prolongation formed of three stars.

These three last stars belong to Andromeda, and themselves abut on Perseus. The last star in the Square of Pegasus is also the first in Andromeda.

[gamma] of Andromeda is a magnificent double orb, to which we shall return in the next chapter, i.e., the telescope resolves it into two marvelous suns, one of which is topaz-yellow, and the other emerald-green. Three stars, indeed, are visible with more powerful instruments.

Above [beta] and near a small star, is visible a faint, whitish, luminous trail: this is the oblong nebula of Andromeda, the first mentioned in the history of astronomy, and one of the most beautiful in the Heavens, perceptible to the unaided eye on very clear nights.

The stars [alpha], [beta] and [gamma] of Perseus form a concave bow which will serve in a new orientation. If it is prolonged in the direction of [delta], we find a very brilliant star of the first magnitude. This is Capella, the Goat, in the constellation of the Charioteer (Fig. 7).

If coming back to [delta] in Perseus, a line is drawn toward the South, we reach the Pleiades, a gorgeous cluster of stars, scintillating like the finest dust of diamonds, on the shoulder of the Bull, to which we shall come shortly, in studying the Constellations of the Zodiac.

Not far off is a very curious star, [beta] of Perseus, or Algol, which forms a little triangle with two others smaller than itself. This star is peculiar in that, instead of shining with a fixed light, it varies in intensity, and is sometimes pale, sometimes brilliant. It belongs to the category of variable stars which we shall study later on. All the observations made on it for more than two hundred years go to prove that a dark star revolves round this sun, almost in the plane of our line of sight, producing as it passes in front of it a partial eclipse that reduces it from the second to the fourth magnitude, every other two days, twenty hours, and forty-nine minutes.

And now, let us return to the Great Bear, which aided us so beneficently to start for these distant shores, and whence we shall set out afresh in search of other constellations.

If we produce the curved line of the tail, or handle, we encounter a magnificent golden-yellow star, a splendid sun of dazzling brilliancy: let us make our bow to Arcturus, [alpha] of the Herdsman, which is at the extremity of this pentagonal constellation. The principal stars of this asterism are of the third magnitude, with the exception of [alpha], which is of the first. Alongside of the Herdsman is a circle consisting of five stars of the third and fourth magnitude, save the third, [alpha], or the Pearl, which is of the second magnitude. This is the Corona Borealis. It is very easily recognized (Fig. 8).

A line drawn from the Pole-Star to Arcturus forms the base of an equilateral triangle, the apex of which, situated opposite the Great Bear, is occupied by Vega, or [alpha] of the Lyre, a splendid diamond of ideal purity scintillating through the ether. This magnificent star, of first magnitude, is, with Arcturus, the most luminous in our Heavens. It burns with a white light, in the proximity of the Milky Way, not far from a constellation that is very easily recognized by the arrangement of its principal stars in the form of a cross. It is named Cygnus, the Bird, or the Swan (Fig. 9), and is easy to find by the Square of Pegasus, and the Milky Way. This figure, the brilliancy of whose constituents (of the third and fourth magnitudes) contrasts strongly with the pallor of the Milky Way, includes at its extremity at the foot of the Cross, a superb double star, [beta] or Albirio: [alpha] of Cygnus is also called Deneb. The first star of which the distance was calculated is in this constellation. This little orb of fifth magnitude, which hangs 69,000,000,000,000 kilometers (42,000,000,000,000 miles) above our Earth, is the nearest of all the stars to the skies of Europe.

Not far off is the fine Eagle, which spreads its wings in the Milky Way, and in which the star Alta [alpha], of first magnitude, is situated between its two satellites, [beta] and [gamma].

The Constellation of Hercules, toward which the motions of the Sun are impelling us, with all the planets of its system, is near the Lyre. Its principal stars can be recognized inside the triangle formed by the Pole-Star, Arcturus, and Vega.

All the Constellations described above belong to the Northern Hemisphere. Those nearest the pole are called circumpolar. They revolve round the pole in twenty-four hours.

Having now learned the Northern Heavens, we must come back to the Sun, which we have left behind us. The Earth revolves round him in a year, and in consequence he seems to revolve round us, sweeping through a vast circle of the celestial sphere. In each year, at the same period, he passes the same points of the Heavens, in front of the same constellations, which are rendered invisible by his light. We know that the stars are at a fixed position from the Earth, whatever their distance, and that if we do not see them at noon as at midnight, it is simply because they are extinguished by the dazzling light of the orb of day. With the aid of a telescope it is always possible to see the more brilliant of them.

The Zodiac is the zone of stars traversed by the Sun in the course of a year. This word is derived from the Greek Zodiakos, which signifies "animal," and this etymology arose because most of the figures traced on this belt of stars represent animals. The belt is divided into twelve parts that are called the twelve Signs of the Zodiac, also named by the ancients the "Houses of the Sun," since the Sun visits one of them in each month. These are the signs, with the primitive characters that distinguish them: the Ram [Aries], the Bull [Taurus], the Twins [Gemini], the Crab [Cancer], the Lion [Leo], the Virgin [Virgo], the Balance [Libra], the Scorpion [Scorpio], the Archer [Sagittarius], the Goat [Capricorn], the Water-Carrier [Aquarius], the Fishes [Pisces]. The sign [Aries] represents the horns of the Ram, [Taurus] the head of the Bull, and so on.

If you will now follow me into the Houses of the Sun you will readily recognize them again, provided you have a clear picture of the principal stars of the Northern Heavens. First, you see the Ram, the initial sign of the Zodiac; because at the epoch at which the actual Zodiac was fixed, the Sun entered this sign at the vernal equinox, and the equator crossed the ecliptic at this point. This constellation, in which the horns of the Ram (third magnitude) are the brightest, is situated between Andromeda and the Pleiades. Two thousand years ago, the Ram was regarded as the symbol of spring; but owing to the secular movement of the precession of the equinoxes, the Sun is no longer there on March 21: he is in the Fishes.

To the left, or east of the Ram, we find the Bull, the head of which forms a triangle in which burns Aldebaran, of first magnitude, a magnificent red star that marks the right eye; and the Hyades, scintillating pale and trembling, on its forehead. The timid Pleiades, as we have seen, veil themselves on the shoulder of the Bull--a captivating cluster, of which six stars can be counted with the unaided eye, while several hundred are discovered with the telescope.

Next the Twins. They are easily recognized by the two fine stars, [alpha] and [beta], of first magnitude, which mark their heads, and immortalize Castor and Pollux, the sons of Jupiter, celebrated for their indissoluble friendship.

Cancer, the Crab, is the least important sign of the Zodiac. It is distinguished only by five stars of fourth and fifth magnitudes, situated below the line of Castor and Pollux, and by a pale cluster called Pr 鍟 epe, the Beehive.

The Lion next approaches, superb in his majesty. At his heart is a gorgeous star of first magnitude, [alpha] or Regulus. This figure forms a grand trapezium of four stars on the celestial sphere.

The Virgin exhibits a splendid star of first magnitude; this is Spica, which with Regulus and Arcturus, form a triangle by which this constellation can be recognized.

The Balance follows the Virgin. Its scales, marked by two stars of second magnitude, are situated a little to the East of Spica.

We next come to the eighth constellation of the Zodiac, which is one of the most beautiful of this belt of stars. Antares, a red star of first magnitude, occupies the heart of the venomous and accursed Scorpion. It is situated on the prolongation of a line joining Regulus to Spica, and forms with Vega of the Lyre, and Arcturus of the Herdsman, a great isosceles triangle, of which this latter star is the apex.

The Scorpion, held to be a sign of ill luck, has been prejudicial to the Archer, which follows it, and traces an oblique trapezium in the sky, a little to the east of Antares. These two southernmost constellations never rise much above the horizon for France and England. In fable, the Archer is Chiron, the preceptor of Jason and Achilles.

Capricorn lies to the south of Alta, on the prolongation of a line from the Lyre to the Eagle. It is hardly noticeable save for the stars [alpha] and [beta] of third magnitude, which scintillate on its forehead.

The Water-Carrier pours his streams toward the horizon. He is not rich in stars, exhibiting only three of third magnitude that form a very flattened triangle.

Lastly the Fishes, concluding sign of the Zodiac, are found to the south of Andromeda and Pegasus. Save for [alpha], of third magnitude, this constellation consists of small stars that are hardly visible.

These twelve zodiacal constellations will be recognized on examining the chart (Figs. 10-11).

We must now visit the stars of the Southern Heavens, some of which are equally deserving of admiration.

It should in the first place be noted that the signs of the Zodiac and the Southern Constellations are not, like those which are circumpolar, perpetually visible at all periods of the year. Their visibility depends on the time of year and the hour of the night.[4]

In order to admire the fine constellations of the North, as described above, we have only to open our windows on a clear summer's evening, or walk round the garden in the mysterious light of these inaccessible suns, while we look up at the immense fields in which each star is like the head of a celestial spear.

But the summer is over, autumn is upon us, and then, too soon, comes winter clothed in hoar-frost. The days are short and cold, dark and dreary; but as a compensation the night is much longer, and adorns herself with her most beautiful jewels, offering us the contemplation of her inexhaustible treasures.

First, let us do homage to the magnificent Orion, most splendid of all the constellations: he advances like a colossal giant, and confronts the Bull.

This constellation appears about midnight in November, in the south-eastern Heavens; toward eleven o'clock in December and January, due south; about ten in February, in the south-east; about nine in March, and about eight in April, in the west; and then sets below our horizon.

It is indisputably the most striking figure in the sky, and with the Great Bear, the most ancient in history, the first that was noticed: both are referred to in the ancient texts of China, Chaldea, and Egypt.

Eight principal stars delineate its outline; two are of the first magnitude, five of the second, and one of the third (Fig. 12). The most brilliant are Betelgeuse ([alpha]) and Rigel ([beta]): the former marking the right shoulder of the Colossus as it faces us; the second the left foot. The star on the left shoulder is [gamma] or Bellatrix, of second magnitude; that of the right foot, [chi], is almost of the third. Three stars of second magnitude placed obliquely at equal distances from each other, the first or highest of which marks the position of the equatorial line, indicate the Belt or Girdle. These stars, known as the Three Kings, and by country people as the Rake, assist greatly in the recognition of this fine constellation.

A little below the second star of the Belt, a large white patch, like a band of fog, the apparent dimensions of which are equal to that of the lunar disk, is visible to the unaided eye: this is the Nebula of Orion, one of the most magnificent in the entire Heavens. It was discovered in 1656 by Huyghens, who counted twelve stars in the pale cloud. Since that date it has been constantly studied and photographed by its many admirers, while the giant eye of the telescope discovers in it to-day an innumerable multitude of little stars which reveal the existence of an entire universe in this region.

Orion is not merely the most imposing of the celestial figures; it is also the richest in sidereal wonders. Among these, it exhibits the most complex of all the multiple systems known to us: that of the star [theta] situated in the celebrated nebula just mentioned. This marvelous star, viewed through a powerful telescope, breaks up into six suns, forming a most remarkable stellar group.

This region is altogether one of the most brilliant in the entire firmament. We must no longer postpone our homage to the brightest star in the sky, the magnificent Sirius, which shines on the left below Orion: it returns every year toward the end of November. This marvelous star, of dazzling brilliancy, is the first, [alpha], in the constellation of the Great Dog, which forms a quadrilateral, the base of which is adjacent to a triangle erected from the horizon.

When astronomers first endeavored to determine the distance of the stars, Sirius, which attracted all eyes to its burning fires, was the particular object of attention. After long observation, they succeeded in determining its distance as 92 trillion kilometers (57 trillion miles). Light, that radiates through space at a velocity of 300,000 kilometers (186,000 miles) per second, takes no less than ten years to reach us from this sun, which, nevertheless, is one of our neighbors.

The Little Dog, in which Procyon ([alpha], of first magnitude) shines out, is above its big brother. With the exception of [alpha], it has no bright stars.

Lastly, toward the southern horizon, we must notice the Hydra, Eridanus, the Whale, the Southern Fish, the Ship, and the Centaur. This last constellation, while invisible to our latitudes, contains the star that is nearest to the Earth, [alpha], of first magnitude, the distance of which is 40 trillion kilometers (25 trillion miles).

The feet of the Centaur touch the Southern Cross, which is always invisible to us, and a little farther down the Southern Pole reigns over the icy desert of the antarctic regions.

In order to complete the preceding descriptions, we subjoin four charts representing the aspect of the starry heavens during the evenings of winter, spring, summer, and autumn. To make use of these, we must suppose them to be placed above our heads, the center marking the zenith, and the sky descending all round to the horizon. The horizon, therefore, bounds these panoramas. Turning the chart in any direction, and looking at it from north, south, east, or west, we find all the principal stars. The first map (Fig. 13) represents the sky in winter (January) at 8 P.M.; the second, in spring (April) at 9 P.M.; the third, in summer (July) at the same hour; the fourth, the sky in autumn (October) at the same time.

And so, at little cost, we have made one of the grandest and most beautiful journeys conceivable. We now have a new country, or, better, have learned to see and know our own country, for since the Earth is a planet we must all be citizens of the Heavens before we can belong to such or such a nation of our lilliputian world.

We must now study this sublime spectacle of the Heavens in detail.

CHAPTER III

THE STARS, SUNS OF THE INFINITE

A JOURNEY THROUGH SPACE

We have seen from the foregoing summary of the principal Constellations that there is great diversity in the brightness of the stars, and that while our eyes are dazzled with the brilliancy of certain orbs, others, on the contrary, sparkle modestly in the azure depths of the night, and are hardly perceptible to the eye that seeks to plumb the abysses of Immensity.

We have appended the word "magnitude" to the names of certain stars, and the reader might imagine this to bear some relation to the volume of the orb. But this is not the case.

To facilitate the observation of stars of varying brilliancy, they have been classified in order of magnitude, according to their apparent brightness, and since the dimensions of these distant suns are almost wholly unknown to us, the most luminous stars were naturally denoted as of first magnitude, those which were a little less bright of the second, and so on. But in reality this word "magnitude" is quite erroneous, for it bears no relation to the mass of the stars, divided thus at an epoch when it was supposed that the most brilliant must be the largest. It simply indicates the apparent brightness of a star, the real brilliancy depending on its dimensions, its intrinsic light, and its distance from our planet.

And now to make some comparison between the different orders. Throughout the entire firmament, only nineteen stars of first magnitude are discoverable. And, strictly speaking, the last of this series might just as well be noted of "second magnitude," while the first of the second series might be added to the list of stars of the "first order." But in order to form classes distinct from one another, some limit has to be adopted, and it was determined that the first series should include only the following stars, the most luminous in the Heavens, which are subjoined in order of decreasing brilliancy.

STARS OF THE FIRST MAGNITUDE

1. Sirius, or [alpha] of the Great Dog. 2. Canopus, or [alpha] of the Ship. 3. Capella, or [alpha] of the Charioteer. 4. Arcturus, or [alpha] of the Herdsman. 5. Vega, or [alpha] of the Lyre. 6. Proxima, or [alpha] of the Centaur. 7. Rigel, or [beta] of Orion. 8. Achernar, or [alpha] of Eridanus. 9. Procyon, or [alpha] of the Little Dog. 10. [beta] of the Centaur. 11. Betelgeuse, or [alpha] of Orion. 12. Alta, or [alpha] of the Eagle. 13. [alpha] of the Southern Cross. 14. Aldebaran, or [alpha] of the Bull. 15. Spica, or [alpha] of the Virgin. 16. Antares, or [alpha] of the Scorpion. 17. Pollux, or [beta] of the Twins. 18. Regulus, or [alpha] of the Lion. 19. Fomalhaut, or [alpha] of the Southern Fish.

THE STARS OF THE SECOND MAGNITUDE

Then come the stars of the second magnitude, of which there are fifty-nine. The stars of the Great Bear (with the exception of [delta], which is of third magnitude), the Pole-Star, the chief stars in Orion (after Rigel and Betelgeuse), of the Lion, of Pegasus, of Andromeda, of Cassiopeia, are of this order. These, with the former, constitute the principal outlines of the constellations visible to us.

Then follow the third and fourth magnitudes, and so on.

* * * * *

The following table gives a summary of the series, down to the sixth magnitude, which is the limit of visibility for the unaided human eye:

19 stars of first magnitude. 59 of second magnitude. 182 of third magnitude. 530 of fourth magnitude. 1,600 of fifth magnitude. 4,800 of sixth magnitude.

This makes a total of some seven thousand stars visible to the unaided eye. It will be seen that each series is, roughly speaking, three times as populated as that preceding it; consequently, if we multiply the number of any class by three, we obtain the approximate number of stars that make up the class succeeding it.

Seven thousand stars! It is an imposing figure, when one reflects that all these lucid points are suns, as enormous as they are potent, as incandescent as our own (which exceeds the volume of the Earth by more than a million times), distant centers of light and heat, exerting their attraction on unknown systems. And yet it is generally imagined that millions of stars are visible in the firmament. This is an illusion; even the best vision is unable to distinguish stars below the sixth magnitude, and ordinary sight is far from discovering all of these.

Again, seven thousand stars for the whole Heavens makes only three thousand five hundred for half the sky. And we can only see one celestial hemisphere at a time. Moreover, toward the horizon, the vapor of the atmosphere veils the little stars of sixth magnitude. In reality, we never see at a given moment more than three thousand stars. This number is below that of the population of a small town.

* * * * *

But celestial space is unlimited, and we must not suppose that these seven thousand stars that fascinate our eyes and enrich our Heavens, without which our nights would be black, dark, and empty,[5] comprise the whole of Creation. They only represent the vestibule of the temple.

Where our vision is arrested, a larger, more powerful eye, that is developing from century to century, plunges its analyzing gaze into the abysses, and reflects back to the insatiable curiosity of science the light of the innumerable suns that it discovers. This eye is the lens of the optical instruments. Even opera-glasses disclose stars of the seventh magnitude. A small astronomical objective penetrates to the eighth and ninth orders. More powerful instruments attain the tenth. The Heavens are progressively transformed to the eye of the astronomer, and soon he is able to reckon hundreds of thousands of orbs in the night. The evolution continues, the power of the instrument is developed; and the stars of the eleventh and twelfth magnitudes are discovered successively, and together number four millions. Then follow the thirteenth, fourteenth, and fifteenth magnitudes. This is the sequence:

7th magnitude 13,000. 8th " 40,000. 9th " 120,000. 10th " 380,000. 11th " 1,000,000. 12th " 3,000,000. 13th " 9,000,000. 14th " 27,000,000. 15th " 80,000,000.

Accordingly, the most powerful telescopes of the day, reenforced by celestial photography, can bring a stream of more than 120 millions of stars into the scope of our vision.

The photographic map of the Heavens now being executed comprises the first fourteen magnitudes, and will give the precise position of some 40,000,000 stars, distributed over 22,054 sheets, forming a sphere 3 meters 44 centimeters in diameter.

The boldest imagination is overwhelmed by these figures, and fails to picture such millions of suns--formidable and burning globes that roll through space, sweeping their systems along with them. What furnaces are there! what unknown lives! what vast immensities!

And again, what enormous distances must separate the stars, to admit of their free revolution in the ether! In what abysses, at what a distance from our terrestrial atom, must these magnificent and dazzling Suns pursue the paths traced for them by Destiny!

* * * * *

If all the stars radiated an equal light, their distances might be calculated on the principle that an object appears smaller in proportion to its distance. But this equality does not exist. The suns were not all cast in the same mold.

Indeed, the stars differ widely in size and brightness, and the distances that have been measured show that the most brilliant are not the nearest. They are scattered through Space at all distances.

Among the nearer stars of which it has been found possible to calculate the distance, some are found to be of the fourth, fifth, sixth, seventh, eighth, and even ninth magnitudes, proving that the most brilliant are not always the least distant.

For the rest, among the beautiful and shining stars with which we made acquaintance in the last chapter may be cited Sirius, which at a distance of 92 trillion kilometers (57 trillion miles) from here still dazzles us with its burning fires; Procyon or [alpha] of the Little Dog, as remote as 112 trillion kilometers (69-1/2 trillion miles); Alta 飑 of the Eagle, at 160 trillion kilometers (99 trillion miles); the white Vega, at 204 trillion kilometers (126-1/2 trillion miles); Capella, at 276 trillion kilometers (171 trillion miles); and the Pole-Star at 344 trillion kilometers (213-1/2 trillion miles). The light that flies through Space at a velocity of 300,000 kilometers (186,000 miles) per second, takes thirty-six years and a half to reach us from this distant sun: i.e., the luminous ray we are now receiving from Polaris has been traveling for more than the third of a century. When you, gentle reader, were born, the ray that arrives to-day from the Pole-Star was already speeding on its way. In the first second after it had started it traveled 300,000 kilometers; in the second it added another 300,000 which at once makes 600,000 kilometers; add another 300,000 kilometers for the third second, and so on during the thirty-six years and a half.

If we tried to arrange the number 300,000 (which represents the distance accomplished in one second) in superposed rows, as if for an addition sum, as many times as is necessary to obtain the distance that separates the Pole-Star from our Earth, the necessary operation would comprise 1,151,064,000 rows, and the sheet of paper required for the setting out of such a sum would measure approximately 11,510 kilometers (about 7,000 miles), i.e., almost the diameter of our terrestrial globe, or about four times the distance from Paris to Moscow!

Is it not impossible to realize that our Sun, with its entire system, is lost in the Heavens at such a distance from his peers in Space? At the distance of the least remote of the stars he would appear as one of the smallest.

* * * * *

The nearest star to us is [alpha] of the Centaur, of first magnitude, a neighbor of the South Pole, invisible in our latitudes. Its distance is 275,000 radii of the terrestrial orbit, i.e., 275,000 times 149 million kilometers, which gives 41 trillions, or 41,000 milliards of kilometers (= 25-1/2 trillion miles). [A milliard = 1,000 millions, the French billion. A trillion = 1,000 milliards, or a million millions, the English billion. The French nomenclature has been retained by the translator.] At a speed of 300,000 kilometers (186,000 miles) per second the light takes four years to come from thence. It is a fine double star.

The next nearest star after this is a little orb invisible to the unaided eye. It has no name, and stands as No. 21,185 in the Catalogue of Lalande. It almost attains the seventh magnitude (6.8). Its distance is 64 trillion kilometers (39-1/2 trillion miles).

The third of which the distance has been measured is the small star in Cygnus, already referred to in Chapter II, in describing the Constellations. Its distance is 69 trillion kilometers (42-1/2 trillion miles). This, too, is a double star. The light takes seven years to reach us.

As we have seen, the fine stars Sirius, Procyon, Aldebaran, Alta 風, Vega, and Capella are more remote.

Our solar system is thus very isolated in the vastness of Infinitude. The latest known planet of our system, Neptune, performs its revolutions in space at 4 milliards, 470 million kilometers (2,771,400,000 miles) from our Sun. Even this is a respectable distance! But beyond this world, an immense gulf, almost a void abyss, extends to the nearest star, [alpha] of the Centaur. Between Neptune and Centauris there is no star to cheer the black and cold solitude of the immense vacuum. One or two unknown planets, some wandering comets, and swarms of meteors, doubtless traverse those unknown spaces, but all invisible to us.

Later on we will discuss the methods that have been employed in measuring these distances. Let us now continue our description.

* * * * *

Now that we have some notion of the distance of the stars we must approach them with the telescope, and compare them one with another.

Let us, for example, get close to Sirius: in this star we admire a sun that is several times heavier than our own, and of much greater mass, accompanied by a second sun that revolves round it in fifty years. Its light is exceedingly white, and it notably burns with hydrogen flames, like Vega and Alta 飏.

Now let us approach Arcturus, Capella, Aldebaran: these are yellow stars with golden rays, like our Sun, and the vapor of iron, of sodium, and of many other metals can be identified in their spectrum. These stars are older than the first, and the ruddy ones, such as Antares, Betelgeuse, [alpha] of Hercules, are still older; several of them are variable, and are on their way to final extinction.

The Heavens afford us a perennial store of treasure, wherein the thinker, poet or artist can find inexhaustible subjects of contemplation.

You have heard of the celestial jewels, the diamonds, rubies, emeralds, sapphires, topazes, and other precious stones of the sidereal casket. These marvels are met with especially among the double stars.

Our Sun, white and solitary, gives no idea of the real aspect of some of its brothers in Infinitude. There are as many different types as there are suns!

Stars, you will think, are like individuals: each has its distinct characteristics: no two are comparable. And indeed this reflection is justified. While human vanity does homage to Phoebus, divine King of the Heavens, other suns of still greater magnificence form groups of two or three splendid orbs, which roll the prodigious combinations of their double, triple, or multiple systems through space, pouring on to the worlds that accompany them a flood of changing light, now blue, now red, now violet, etc.

In the inexhaustible variety of Creation there exist Suns that are united in pairs, bound by a common destiny, cradled in the same attraction, and often colored in the most delicate and entrancing shades conceivable. Here will be a dazzling ruby, its glowing color shedding joy; there a deep blue sapphire of tender tone; beyond, the finest emeralds, hue of hope. Diamonds of translucent purity and whiteness sparkle from the abyss, and shed their penetrating light into the vast space. What splendors are scattered broadcast over the sky! what profusion!

To the naked eye, the groups appear like ordinary stars, mere luminous points of greater or less brilliancy; but the telescope soon discovers the beauty of these systems; the star is duplicated into two distinct suns, in close proximity. These groups of two or several suns are not merely due to an effect of perspective--i.e., the presence of two or more stars in our line of sight; as a rule they constitute real physical systems, and these suns, associated in a common lot, rotate round one another in a more or less rapid period, that varies for each system.

One of the most splendid of these double stars, and at the same time one of the easiest to perceive, is [zeta] in the Great Bear, or Mizar, mentioned above in describing this constellation. It has no contrasting colors, but exactly resembles twin diamonds of the finest water, which fascinate the gaze, even through a small objective.

Its components are of the second and fourth magnitudes, their distance = 14"[6]. Some idea of their appearance in a small telescope may be obtained from the subjoined figure (Fig. 17).

Another very brilliant pair is Castor. Magnitudes second and third. Distance 5.6"". Very easy to observe. [gamma] in the Virgin resolves into two splendid diamonds of third magnitude. Distance, 5.0". Another double star is [gamma] of the Ram, of fourth magnitude. Distance, 8.9".

And here are two that are even more curious by reason of their coloring: [gamma] in Andromeda, composed of a fine orange star, and one emerald-green, which again is accompanied by a tiny comrade of the deepest blue. This group in a good telescope is most attractive. Magnitudes, second and fifth. Distance, 10".

[beta] of the Swan, or Albireo, referred to in the last chapter, has been analyzed into two stars: one golden-yellow, the other sapphire. Magnitudes, third and fifth. Distance, 34". [alpha] of the Greyhounds, known also as the Heart of Charles II, is golden-yellow and lilac. Magnitudes, third and fifth. Distance 20".[7]

[alpha] of Hercules revolves a splendid emerald and a ruby in the skies; [zeta] of the Lyre exhibits a yellow and a green star; Rigel, an electric sun, and a small sapphire; Antares is ruddy and emerald-green; [eta] of Perseus resolves into a burning red star, and one smaller that is deep blue, and so on.

* * * * *

These exquisite double stars revolve in gracious and splendid couples around one another, as in some majestic valse, marrying their multi-colored fires in the midst of the starry firmament.

Here, we constantly receive a pure and dazzling white light from our burning luminary. Its ray, indeed, contains the potentiality of every conceivable color, but picture the fantastic illumination of the worlds that gravitate round these multiple and colored suns as they shed floods of blue and roseate, red, or orange light around them! What a fairy spectacle must life present upon these distant universes!

Let us suppose that we inhabit a planet illuminated by two suns, one blue, the other red.

It is morning. The sapphire sun climbs slowly up the Heavens, coloring the atmosphere with a somber and almost melancholy hue. The blue disk attains the zenith, and is beginning its descent toward the West, when the East lights up with the flames of a scarlet sun, which in its turn ascends the heights of the firmament. The West is plunged in the penumbra of the rays of the blue sun, while the East is illuminated with the purple and burning rays of the ruby orb.

The first sun is setting when the second noon shines for the inhabitants of this strange world. But the red sun, too, accomplishes the law of its destiny. Hardly has it disappeared in the conflagration of its last rays, with which the West is flushed, when the blue orb reappears on the opposite side, shedding a pale azure light upon the world it illuminates, which knows no night. And thus these two suns fraternize in the Heavens over the common task of renewing a thousand effects of extra-terrestrial light for the globes that are subject to their variations.

Scarlet, indigo, green, and golden suns; pearly and multi-colored Moons; are these not fairy visions, dazzling to our poor sight, condemned while here below to see and know but one white Sun?

As we have learned, there are not only double, but triple, and also multiple stars. One of the finest ternary systems is that of [gamma] in Andromeda, above mentioned. Its large star is orange, its second green, its third blue, but the two last are in close juxtaposition, and a powerful telescope is needed to separate them. A triple star more easy to observe is [zeta] of Cancer, composed of three orbs of fifth magnitude, at a distance of 1" and 5"; the first two revolve round their common center of gravity in fifty-nine years, the third takes over three hundred years. The preceding figure shows this system in a fairly powerful objective (Fig. 18).

In the Lyre, a little above the dazzling Vega, [epsilon] is of fourth magnitude, which seems a little elongated to the unaided eye, and can even be analyzed into two contiguous stars by very sharp sight. But on examining this attractive pair with a small glass, it is further obvious that each of these stars is double; so that they form a splendid quadruple system of two couples (Fig. 19): one of fifth and a half and sixth magnitudes, at a distance of 2.4", the other of sixth and seventh, 3.2" distant. The distance between the two pairs is 207".

In speaking of Orion, we referred to the marvelous star [theta] situated in the no less famous Nebula, below the Belt; this star forms a dazzling sextuple system, in the very heart of the nebula (Fig. 20). How different to our Sun, sailing through Space in modest isolation!

Be it noted that all these stars are animated by prodigious motions that impel them in every direction.

There are no fixed stars. On every side throughout Infinity, the burning suns-- enormous globes, blazing centers of light and heat--are flying at giddy speed toward an unknown goal, traversing millions of miles each day, crossing century by century such vast spaces as are inconceivable to the human intellect.

If the stars appear motionless to us, it is because they are so remote, their secular movements being only manifested on the celestial sphere by imperceptible displacements. But in reality these suns are in perpetual commotion in the abysses of the Heavens, which they quicken with an extraordinary animation.

These perpetual and cumulative motions must eventually modify the aspect of the Constellations: but these changes will only take effect very slowly; and for thousands and thousands of years longer the heroes and heroines of mythology will keep their respective places in the Heavens, and reign undisturbed beneath the starry vault.

Examination of these star motions reveals the fact that our Sun is plunging with all his system (the Earth included) toward the Constellation of Hercules. We are changing our position every moment: in an hour we shall be 70,000 kilometers (43,500 miles) farther than we are at present. The Sun and the Earth will never again traverse the space they have just left, and which they have deserted forever.

And here let us pause for an instant to consider the variable stars. Our Sun, which is constant and uniform in its light, does not set the type of all the stars. A great number of them are variable--either periodically, in regular cycles--or irregularly.

We are already acquainted with the variations of Algol, in Perseus, due to its partial eclipse by a dark globe gravitating in the line of our vision. There are several others of the same type: these are not, properly speaking, variable stars. But there are many others the intrinsic light of which undergoes actual variations.

In order to realize this, let us imagine that our Earth belongs to such a sun, for example, to a star in the southern constellation of the Whale, indicated by the letter [omicron], which has been named the "wonderful" (Mira Ceti). Our new sun is shining to-day with a dazzling light, shedding the gladness of his joyous beams upon nature and in our hearts. For two months we admire the superb orb, sparkling in the azure illuminated with its radiance. Then of a sudden, its light fades, and diminishes in intensity, though the sky remains clear. Imperceptibly, our fine sun darkens; the atmosphere becomes sad and dull, there is an anticipation of universal death. For five long months our world is plunged in a kind of penumbra; all nature is saddened in the general woe.

But while we are bewailing the cruelty of our lot, our cherished luminary revives. The intensity of its light increases slowly. Its brilliancy augments, and finally, at the end of three months, it has recovered its former splendors, and showers its bright beams upon our world, flooding it with joy. But--we must not rejoice too quickly! This splendid blaze will not endure. The flaming star will pale once more; fade back to its minimum; and then again revive. Such is the nature of this capricious sun. It varies in three hundred and thirty-one days, and from yellow at the maximum, turns red at the minimum. This star, Mira Ceti, which is one of the most curious of its type, varies from the second to the ninth magnitudes: we cite it as one example; hundreds of others might be instanced.

Thus the sky is no black curtain dotted with brilliant points, no empty desert, silent and monotonous. It is a prodigious theater on which the most fantastic plays are continually being acted. Only--there are no spectators.

Again, we must note the temporary stars, which shine for a certain time, and then die out rapidly. Such was the star in Cassiopeia, in 1572, the light of which exceeded Sirius in its visibility in full daylight, burning for five months with unparalleled splendor, dominating all other stars of first magnitude; after which it died out gradually, disappearing at the end of seventeen months, to the terror of the peoples, who saw in it the harbinger of the world's end: that of 1604, in the Constellation of the Serpent, which shone for a year; of 1866, of second magnitude, in the Northern Crown, which appeared for a few weeks only; of 1876, in the Swan; of 1885, in the Nebula of Andromeda; of 1891, in the Charioteer; and quite recently, of 1901, in Perseus.

These temporary stars, which appear spontaneously to the observers on the Earth, and quickly vanish again, are doubtless due to collisions, conflagrations, or celestial cataclysms. But we only see them long after the epoch at which the phenomena occurred, years upon years, and centuries ago. For instance, the conflagration photographed by the author in 1901, in Perseus, must have occurred in the time of Queen Elizabeth. It has taken all this time for the rays of light to reach us.

* * * * *

The Heavens are full of surprises, on which we can bestow but a fleeting glance within these limits. They present a field of infinite variety.

Who has not noticed the Milky Way, the pale belt that traverses the entire firmament and is so luminous on clear evenings in the Constellations of the Swan and the Lyre? It is indeed a swarm of stars. Each is individually too small to excite our retina, but as a whole, curiously enough, they are perfectly visible. With opera-glasses we divine the starry constitution: a small telescope shows us marvels. Eighteen millions of stars were counted there with the gauges of William Herschel.

Now this Milky Way is a symbol, not of the Universe, but of the Universes that succeed each other through the vast spaces to Infinity.

Our Sun is a star of the Milky Way. It surrounds us like a great circle, and if the Earth were transparent, we should see it pass beneath our feet as well as over our heads. It consists of a very considerable mass of star-clusters, varying greatly in extent and number, some projected in front of others, while the whole forms an agglomeration.

Among this mass of star-groups, several thousands of which are already known to us, we will select one of the most curious, the Cluster in Hercules, which can be distinguished with the unaided eye, between the stars [eta] and [zeta] of that constellation. Many photographs of it have been taken in the author's observatory at Juvisy, showing some thousands of stars; and one of these is reproduced in the accompanying figure (Fig. 21). Is it not a veritable universe?

[Illustration: FIG. 22.--The Star-Cluster in the Centaur.]

Another of the most beautiful, on account of its regularity, is that of the Centaur (Fig. 22).

These groups often assume the most extraordinary shapes in the telescope, such as crowns, fishes, crabs, open mouths, birds with outspread wings, etc.

We must also note the gaseous nebul? universes in the making, e.g., the famous Nebula in Orion, of which we obtained some notion a while ago in connection with its sextuple star: and also that in Andromeda (Fig. 23).

Perhaps the most marvelous of all is that of the Greyhounds, which evolves in gigantic spirals round a dazzling focus, and then loses itself far off in the recesses of space. Fig. 24 gives a picture of it.

Without going thus far, and penetrating into telescopic depths, my readers can get some notion of these star-clusters with the help of a small telescope or opera-glasses, or even with the unaided eye, by looking at the beautiful group of the Pleiades, already familiar to us on another page, and using it as a test of vision. The little map subjoined (Fig. 25) will be an assistance in recognizing them, and in estimating their magnitudes, which are in the following order:

Alcyone 3.0. Electra 4.5. Atlas 4.6. Maia 5.0. Merope 5.5. Taygeta 5.8. Pleione 6.3. Celio 6.5. Asterope 6.8.

Good eyes distinguish the first six, sharp sight detects the three others.

In the times of the ancient Greeks, seven were accounted of equal brilliancy, and the poets related that the seventh star had fled at the time of the Trojan War. Ovid adds that she was mortified at not being embraced by a god, as were her six sisters. It is probable that only the best sight could then distinguish Pleione, as in our own day. The angular distance from Atlas to Pleione is 5'.

The length of this republic, from Atlas and Pleione to Celio, is 4'/23" of time, or 1?' of arc; the breadth, from Merope to Asterope, is 36'.[8]

In the quadrilateral, the length from Alcyone to Electra is 36', and the breadth from Merope to Maia 25'. To us it appears as though, if the Full Moon were placed in front of this group of nine stars, she would cover it entirely, for to the naked eye she appears much larger than all the Pleiades together. But this is not so. She only measures 31', less than half the distance from Atlas to Celio; she is hardly broader than the distance from Alcyone to Atlas, and could pass between Merope and Taygeta without touching either of these stars. This is a perennial and very curious optical illusion. When the Moon passes in front of the Pleiades, and occults them successively, it is hard to believe one's eyes. The fact occurred, e.g., on July 23, 1897, during a fine occultation observed at the author's laboratory of Juvisy (Fig. 26).

Photography here discovers to us, not 6, 9, 12, 15, or 20 stars, but hundreds and millions.

These are the most brilliant flowers of the celestial garden.

We, alas, can but glance at them rapidly. In contemplating them we are transported into immensities both of space and time, for the stellar periods measured by these distant universes often overpower in their magnitude the rapid years in which our terrestrial days are estimated. For instance, one of the double stars we spoke of above, [gamma] of the Virgin, sees its two components, translucent diamonds, revolve around their common center of gravity, in one hundred and eighty years. How many events took place in France, let us say, in a single year of this star!--The Regency, Louis XV, Louis XVI, the Revolution, Napoleon, Louis XVIII, Louis Philippe, the Second Republic, Napoleon III, the Franco-German War, the Third Republic.... What revolutions here, during a single year of this radiant pair! (Fig. 27.)

But the pageant of the Heavens is too vast, too overwhelming. We must end our survey.

Our Milky Way, with its millions of stars, represents for us only a portion of the Creation. The illimitable abysses of Infinitude are peopled by other universes as vast, as imposing, as our own, which are renewed in all directions through the depths of Space to endless distance. Where is our little Earth? Where our Solar System? We are fain to fold our wings, and return from the Immense and Infinite to our floating island.

CHAPTER IV

OUR STAR THE SUN

In the incessant agitation of daily life in which we are involved by the thousand superfluous wants of modern "civilization," one is prone to assume that existence is complete only when it reckons to the good an incalculable number of petty incidents, each more insignificant than the last. Why lose time in thinking or dreaming? We must live at fever heat, must agitate, and be infatuated for inanities, must create imaginary desires and torments.

The thoughtful mind, prone to contemplation and admiration of the beauties of Nature, is ill at ease in this perpetual vortex that swallows everything--satisfaction, in a life that one has not time to relish; love of the beautiful, that one views with indifference; it is a whirlpool that perpetually hides Truth from us, forgotten forever at the bottom of her well.

And why are our lives thus absorbed in merely material interests? To satisfy our pride and vanity! To make ourselves slaves to chimeras! If the Moon were inhabited, and if her denizens could see us plainly enough to note and analyze the details of human existence on the surface of our planet, it would be curious and perhaps a little humiliating for us, to see their statistics. What! we should say, is this the sum of our lives? Is it for this that we struggle, and suffer, and die? Truly it is futile to give ourselves such trouble.

And yet the remedy is simple, within the power of every one; but one does not think of it just because it is too easy, although it has the immense advantage of lifting us out of the miseries of this weary world toward the inexpressible happiness that must always awaken in us with the knowledge of the Truth: we need only open our eyes to see, and to look out. Only--one hardly ever thinks of it, and it is easier to let one's self be blinded by the illusion and false glamor of appearances.

Think what it would be to consecrate an hour each day to voluntary participation in the harmonious Choir of Nature, to raise one's eyes toward the Heavens, to share the lessons taught by the Pageant of the Universe! But, no: there is no time, no time for the intellectual life, no time to become attached to real interests, no time to pursue them.

Among the objects marshaled for us in the immense spectacle of Nature, nothing without exception has struck the admiration and attention of man as much as the Sun, the God of Light, the fecundating orb, without which our planet and its life would never have issued from nonentity, the visible image of the invisible god, as said Cicero, and the poets of antiquity. And yet how many beyond the circle of those likely to read these pages know that this Sun is a star in the Milky Way, and that every star is a sun? How many take any account of the reality and grandeur of the Universe? Inquire, and you will find that the number of people who have any notion, however rudimentary, of its construction, is singularly restricted. Humanity is content to vegetate, much after the fashion of a race of moles.

Henceforward, you will know that you are living in the rays of a star, which, from its proximity, we term a sun. To the inhabitants of other systems of worlds, our splendid Sun is only a more or less brilliant, luminous point, according as the spot from which it is observed is nearer or farther off. But to us its "terrestrial" importance renders it particularly precious; we forget all the sister stars on its account, and even the most ignorant hail it with enthusiasm without exactly knowing what its r 鬺 e in the universe may be, simply because they feel that they depend on it, and that without it life would become extinct on this globe. Yes, it is the beneficent rays of the Sun that shed upon our Earth the floods of light and heat to which Life owes its existence and its perpetual propagation.

Hail, vast Sun! a little star in Infinitude, but for us a colossal and portentous luminary. Hail, divine Benefactor! How should we not adore, when we owe him the glow of the warm and cheery days of summer, the gentle caresses by which his rays touch the undulating ears, and gild them with the touch? The Sun sustains our globe in Space, and keeps it within his rays by the mysteriously powerful and delicate cords of attraction. It is the Sun that we inhale from the embalmed corollas of the flowers that uplift their gracious heads toward his light, and reflect his splendors back to us. It is the Sun that sparkles in the foam of the merry wine; that charms our gaze in those first days of spring, when the home of the human race is adorned with all the charms of verdant and flowering youth. Everywhere we find the Sun; everywhere we recognize his work, extending from the infinitely great to the infinitely little. We bow to his might, and admire his power. When in the sad winter day he disappears behind the snowy eaves, we think his fiery globe will never rise to mitigate the short December days which are alleviated with his languid beams.

April restores him to superb majesty, and our hearts are filled with hope in the illumination of those beauteous, sunny hours.

* * * * *

Our celestial journey carried us far indeed from our own Solar System. Guided by the penetrating eye of the telescope, we reached such distant creations that we lost sight of our cherished luminary.

But we remember that he burns yonder, in the midst of the pale cosmic cloud we term the Milky Way. Let us approach him, now that we have visited the Isles of Light in the Celestial Ocean; let us traverse the vast plains strewn with the burning gold of the Suns of the Infinite.

We embark upon a ray of light, and glide rapidly to the portals of our Universe. Soon we perceive a tiny speck, scintillating feebly in the depths of Space, and recognize it as our own celestial quarters. This little star shines like the head of a gold pin, and increases in size as we advance toward it. We traverse a few more trillion miles in our rapid course, and it shines out like a fine star of the first magnitude. It grows larger and larger. Soon we divine that it is our humble Earth that is shining before us, and gladly alight upon her. In future we shall not quit our own province of the Celestial Kingdom, but will enter into relations with this solar family, which interests us the more in that it affects us so closely.

The Sun, which is manifested to us as a fine white disk at noon, while it is fiery red in the evening, at its setting, is an immense globe, whose colossal dimensions surpass those of our terrestrial atom beyond all conceivable proportion.

In diameter, it is, in effect, 108-1/2 times as large as the Earth; that is to say, if our planet be represented by a globe 1 meter in diameter, the Sun would figure as a sphere 108-1/2 meters across. This is shown on the accompanying figure (Fig. 28), which is in exact proportion.

If our world were set down upon the Sun, with all its magnificence, all its wealth, its mountains, its seas, its monuments, and its inhabitants, it would only be an imperceptible speck. It would occupy less space in the central orb than one grain in a grenade. If the Earth were placed in the center of the Sun, with the Moon still revolving round it at her proper distance of 384,000 kilometers (238,500 miles), only half the solar surface would be covered.

In volume the Sun is 1,280,000 times vaster than our abode, and 324,000 times heavier in mass. That the giant only appears to us as a small though very brilliant disk, is solely on account of its distance. Its apparent dimensions by no means reveal its majestic proportions to us.

When observed with astronomical instruments, or photographed, we discover that its surface is not smooth, as might be supposed, but granulated, presenting a number of luminous points dispersed over a more somber background. These granulations are somewhat like the pores of a fruit, e.g., a fine orange, the color of which recalls the hue of the Sun when it sinks in the evening, and prepares to plunge us into darkness. At times these pores open under the influence of disturbances that arise upon the solar surface, and give birth to a Sun-Spot. For centuries scientists and lay people alike refused to admit the existence of these spots, regarding them as so many blemishes upon the King of the Heavens. Was not the Sun the emblem of inviolable purity? To find any defect in him were to do him grievous injury. Since the orb of day was incorruptible, those who threw doubt on his immaculate splendor were fools and idiots. And so when Scheiner, one of the first who studied the solar spots with the telescope, published the result of his experiments in 1610, no one would believe his statements.

Yet, from the observations of Galileo and other astronomers, it became necessary to accept the evidence, and stranger still to recognize that it is by these very spots that we are enabled to study the physical constitution of the Sun.

They are generally rounded or oval in shape, and exhibit two distinct parts; first, the central portion, which is black, and is called the nucleus, or umbra; second, a clearer region, half shaded, which has received the name of penumbra. These parts are sharply defined in outline; the penumbra is gray, the nucleus looks black in relation to the dazzling brilliancy of the solar surface; but as a matter of fact it radiates a light 2,000 times superior in intensity to that of the full moon.

Some idea of the aspect of these spots may be obtained from the accompanying reproduction of a photograph of the Sun (taken September 8, 1898, at the author's observatory at Juvisy), and from the detailed drawing of the large spot that broke out some days later (September 13), crossed by a bridge, and furrowed with flames. As a rule, the spots undergo rapid transformations.

These spots, which appear of insignificant dimensions to the observers on the

Earth, are in reality absolutely gigantic. Some that have been measured are ten times as large as the Earth's diameter, i.e., 120,000 kilometers (74,500 miles).

Sometimes the spots are so large that they can be seen with the unaided eye (protected with black or dark-blue glasses). They are not formed instantaneously, but are heralded by a vast commotion on the solar surface, exhibiting, as it were, luminous waves or facul? Out of this agitation arises a little spot, that is usually round, and enlarges progressively to reach a maximum, after which it diminishes, with frequent segmentation and shrinkage. Some are visible only for a few days; others last for months. Some appear, only to be instantly swallowed in the boiling turmoil of the flaming orb. Sometimes, again, white incandescent waves emerge, and seem to throw luminous bridges across the central umbra. As a rule the spots are not very profound. They are funnel-shaped depressions, inferior in depth to the diameter of the Earth, which, as we have seen, is 108 times smaller than that of the Sun.

* * * * *

The Sun-Spots are not devoid of motion, and from their movements we learn that the radiant orb revolves upon itself in about twenty-five days. This rotation was determined in 1611, by Galileo, who, while observing the spots, saw that they traversed the solar disk from east to west, following lines that are oblique to the plane of the ecliptic, and that they disappear at the western border fourteen days after their arrival at the eastern edge. Sometimes the same spot, after being invisible for fourteen days, reappears upon the eastern edge, where it was observed twenty-eight days previously. It progresses toward the center of the Sun, which is reached in seven days, disappears anew in the west, and continues its journey on the hemisphere opposed to us, to reappear under observation two weeks later, if it has not meantime been extinguished. This observation proves that the Sun revolves upon itself. The reappearance of the spots occurs in about twenty-seven days, because the Earth is not stationary, and in its movement round the burning focus, a motion effected in the same direction as the solar rotation, the spots are still visible two and a half days after they disappeared from the point at which they had been twenty-five days previously. In reality, the rotation of the Sun occupies twenty-five and a half days, but strangely enough this globe does not rotate in one uniform period, like the Earth; the rotation periods, or movements of the different parts of the solar surface, diminish from the Sun's equator toward its poles. The period is twenty-five days at the equator, twenty-six at the twenty-fourth degree of latitude, north or south, twenty-seven at the thirty-seventh degree, twenty-eight at the forty-eighth. The spots are usually formed between the equator and this latitude, more especially between the tenth and thirtieth degrees. They have never been seen round the poles.

Toward the edges of the Sun, again, are very brilliant and highly luminous regions, which generally surround the spots, and have been termed facul?(facula, a little torch). These facul? which frequently occupy a very extensive surface, seem to be the seat of formidable commotions that incessantly revolutionize the face of our monarch, often, as we said, preceding the spots. They can be detected right up to the poles.

Our Sun, that appears so calm and majestic, is in reality the seat of fierce conflagrations. Volcanic eruptions, the most appalling storms, the worst cataclysms that sometimes disturb our little world, are gentle zephyrs compared with the solar tempests that engender clouds of fire capable at one burst of engulfing globes of the dimensions of our planet.

To compare terrestrial volcanoes with solar eruptions is like comparing the modest night-light that consumes a midge with the flames of the fire that destroys a town.

The solar spots vary in a fairly regular period of eleven to twelve years. In certain years, e.g., 1893, they are vast, numerous and frequent; in other years, e.g., 1901, they are few and insignificant. The statistics are very carefully preserved. Here, for instance, is the surface showing sun-spots expressed in millionths of the extent of the visible solar surface:

1889 78 1890 99 1891 569 1892 1,214 1893 1,464 1895 974 1896 543 1897 514 1898 375 1899 111 1900 75 1901 29 1902 62

The years 1889 and 1901 were minima; the year 1893 a maximum.

It is a curious fact that terrestrial magnetism and the boreal auroras exhibit an oscillation parallel to that of the solar spots, and apparently the same occurs with regard to temperature.

We must regard our sun as a globe of gas in a state of combustion, burning at high temperature, and giving off a prodigious amount of heat and light. The dazzling surface of this globe is called a photosphere (light sphere). It is in perpetual motion, like the waves of an ocean of fire, whose roseate and transparent flames measure some 15,000 kilometers (9,300 miles) in height. This stratum of rose-colored flames has received the name of chromosphere (color sphere). It is transparent; it is not directly visible, but is seen only during the total eclipses of the Sun, when the dazzling disk of that luminary is entirely concealed by the Moon; or with the aid of the spectroscope. The part of the Sun that we see is its luminous surface, or photosphere.

From this agitated surface there is a constant ejection of gigantic eruptions, immense jets of flame, geysers of fire, projected at a terrific speed to prodigious heights.

For years astronomers were greatly perplexed as to the nature of these incandescent masses, known as prominences, which shot out like fireworks, and were only visible during the total eclipses of the Sun. But now, thanks to an ingenious invention of Janssen and Lockyer, these eruptions can be observed every day in the spectroscope, and have been registered since 1868, more particularly in Rome and in Catania, where the Society of Spectroscopists was founded with this especial object, and publishes monthly bulletins in statistics of the health of the Sun.

These prominences assume all imaginable forms, and often resemble our own storm-clouds; they rise above the chromosphere with incredible velocity, often exceeding 200 kilometers (124 miles) per second, and are carried up to the amazing height of 300,000 kilometers (186,000 miles).

.e., 18 times the diameter of the Earth.]

The Sun is surrounded with these enormous flames on every side; sometimes they shoot out into space like splendid curving roseate plumes; at others they rear their luminous heads in the Heavens, like the cleft and waving leaves of giant palm-trees. Having illustrated a remarkable type of solar spot, it is interesting to submit to the reader a precise observation of these curious solar flames. That reproduced here was observed in Rome, January 30, 1885. It measured 228,000 kilometers (141,500 miles) in height, eighteen times the diameter of the earth (represented alongside in its relative magnitude). (Fig. 31.)

Solar eruptions have been seen to reach, in a few minutes, a height of more than 100,000 kilometers (62,000 miles), and then to fall back in a flaming torrent into that burning and inextinguishable ocean.

Observation, in conjunction with spectral analysis, shows these prominences to be due to formidable explosions produced within the actual substance of the Sun, and projecting masses of incandescent hydrogen into space with considerable force.

Nor is this all. During an eclipse one sees around the black disk of the Moon as it passes in front of the Sun and intercepts its light, a brilliant and rosy aureole with long, luminous, branching feathers streaming out, like aigrettes, which extend a very considerable distance from the solar surface. This aureole, the nature of which is still unknown to us, has received the name of corona. It is a sort of immense atmosphere, extremely rarefied. Our superb torch, accordingly, is a brazier of unparalleled activity--a globe of gas, agitated by phenomenal tempests whose flaming streamers extend afar. The smallest of these flames is so potent that it would swallow up our world at a single breath, like the bombs shot out by Vesuvius, that fall back within the crater.

What now is the real heat of this incandescent focus? The most accurate researches estimate the temperature of the surface of the Sun at 7,000 癈. The internal temperature must be considerably higher. A crucible of molten iron poured out upon the Sun would be as a stream of ice and snow.

We can form some idea of this calorific force by making certain comparisons. Thus, the heat given out appears to be equal to that which would be emitted by a colossal globe of the same dimensions (that is, as voluminous as twelve hundred and eighty thousand terrestrial globes), entirely covered with a layer of incandescent coal 28 kilometers (18 miles) in depth, all burning at equal combustion. The heat emitted by the Sun, at each second, is equal to that which would result from the combustion of eleven quadrillions six hundred thousand milliards of tons of coal, all burning together. This same heat would bring to the boil in an hour, two trillions nine hundred milliards of cubic kilometers of water at freezing-point.

Our little planet, gravitating at 149,000,000 kilometers (93,000,000 miles) from the Sun, arrests on the way, and utilizes, only the half of a milliard part of this total radiation.

How is this heat maintained? One of the principal causes of the heat of the Sun is its condensation. According to all probabilities, the solar globe represents for us the nucleus of a vast nebula, that extended in primitive times beyond the orbit of Neptune, and which in its contraction has finally produced this central focus. In virtue of the law of transformation of motion into heat, this condensation, which has not yet reached its limit, suffices to raise this colossal globe to its level of temperature, and to maintain it there for millions of years. In addition, a substantial number of meteors is forever falling into it. This furnace is a true pandemonium.

The Sun weighs three hundred and twenty-four thousand times more than the Earth--that is to say, eighteen hundred and seventy octillions of kilograms:

1,870,000,000,000,000,000,000,000,000,000
(1,842,364,532,019,704,433,497,536,945 tons).

In Chapter XI we shall explain the methods by which it has been found possible to weigh the Sun and determine its exact distance.

* * * * *

I trust these figures will convey some notion of the importance and nature of the Sun, the stupendous orb on whose rays our very existence depends. Its apparent dimension (which is only half a degree, 32', and would be hidden from sight, like that of the full moon, which is about the same, by the tip of the little finger held out at arm's length), represents, as we have seen, a real dimension that is colossal, i.e., 1,383,000 kilometers (more than 857,000 miles), and this is owing to the enormous distance that separates us from it. This distance of 149,000,000 kilometers (93,000,000 miles) is sufficiently hard to appreciate. Let us say that 11,640 terrestrial globes would be required to throw a bridge from here to the Sun, while 30 would suffice from the Earth to the Moon. The Moon is 388 times nearer to us than the Sun. We may perhaps conceive of this distance by calculating that a train, moving at constant speed of 1 kilometer (0.6214 mile) a minute, would take 149,000,000 minutes, that is to say 103,472 days, or 283 years, to cross the distance that separates us from this orb. Given the normal duration of life, neither the travelers who set out for the Sun, nor their children, nor their grandchildren, would arrive there: only the seventh generation would reach the goal, and only the fourteenth could bring us back news of it.

Children often cry for the Moon. If one of these inquisitive little beings could stretch out its arms to touch the Sun, and burn its fingers there, it would not feel the burn for one hundred and sixty-seven years (when it would no longer be an infant), for the nervous impulse of sensation can only be transmitted from the ends of the fingers to the brain at a velocity of 28 meters per second.

'Tis long. A cannon-ball would reach the Sun in ten years. Light, that rapid arrow that flies through space at a velocity of 300,000 kilometers (186,000 miles per second), takes only eight minutes seventeen seconds to traverse this distance.

* * * * *

This brilliant Sun is not only sovereign of the Earth; he is also the head of a vast planetary system.

The orbs that circle round the Sun are opaque bodies, spherical in shape, receiving their light and heat from the central star, on which they absolutely depend. The name of planets given to them signifies "wandering" stars. If you observe the Heavens on a fine starry night, and are sufficiently acquainted with the principal stars of the Zodiac as described in a preceding chapter, you may be surprised on certain evenings to see the figure of some zodiacal constellation slightly modified by the temporary presence of a brilliant orb perhaps surpassing in its luminosity the finest stars of the first magnitude.

If you watch this apparition for some weeks, and examine its position carefully in regard to the adjacent stars, you will observe that it changes its position more or less slowly in the Heavens. These wandering orbs, or planets, do not shine with intrinsic light; they are illuminated by the Sun.

The planets, in effect, are bodies as opaque as the Earth, traveling round the God of Day at a speed proportional to their distance. They number eight principal orbs, and may be divided into two quite distinct groups by which we may recognize them: the first comprises four planets, of relatively small dimensions in comparison with those of the second group, which are so voluminous that the least important of them is larger than the other four put together.

In order of distance from the Sun, we first encounter:

MERCURY, VENUS, THE EARTH, AND MARS

These are the worlds that are nearest to the orb of day.

The four following, and much more remote, are, still in order of distance:

JUPITER, SATURN, URANUS, AND NEPTUNE

This second group is separated from the first by a vast space occupied by quite a little army of minute planets, tiny cosmic bodies, the largest of which measures little more than 100 kilometers (62 miles) in diameter, and the smallest some few miles only.

The planets which form these three groups represent the principal members of the solar family. But the Sun is a patriarch, and each of his daughters has her own children who, while obeying the paternal influence of the fiery orb, are also obedient to the world that governs them. These secondary asters, or satellites, follow the planets in their course, and revolve round them in an ellipse, just as the others rotate round the Sun. Every one knows the satellite of the Earth, the Moon. All the other planets of our system have their own moons, some being even more favored than ourselves in this respect, and having several. Mars has two; Jupiter, five; Saturn, eight; Uranus, four; and Neptune, one (at least as yet discovered).

In order to realize the relations between these worlds, we must appreciate their distances by arranging them in a little table:

Distance in Distance in Millions of Millions of Kilometers. Miles. Mercury

57 35 Venus 108 67 The Earth 149 93 Mars 226 140 Jupiter 775 481 Saturn 1,421 882 Uranus 2,831 1,755 Neptune 4,470 2,771

The Sun is at the center (or, more properly speaking, at the focus, for the planets describe an ellipse) of this system, and controls them. Neptune is thirty times farther from the Sun than the Earth. These disparities of distance produce a vast difference in the periods of the planetary revolutions; for while the Earth revolves round the Sun in a year, Venus in 224 days, and Mercury in 88, Mars takes nearly 2 years to accomplish his journey, Jupiter 12 years, Saturn 29, Uranus 84, and Neptune 165.

Even the planets and their moons do not represent the Sun's complete paternity. There are further, in the solar republic, certain vagabond and irregular orbs that travel at a speed that is often most immoderate, occasionally approaching the Sun, not to be consumed therein, but, as it appears, to draw from its radiant source the provision of forces necessary for their perigrinations through space. These are the Comets, which pursue an extremely elongated orbit round the Sun, to which at times they approximate very closely, at other times being excessively distant.

And now to recapitulate our knowledge of the Solar Empire. In the first place, we see a colossal globe of fire dominating and governing the worlds that belong to him. Around him are grouped planets, in number eight principal, formed of solid and obscure matter, gravitating round the central orb. Other secondary orbs, the satellites, revolve round the planets, which keep them within the sphere of their attraction. And lastly, the comets, irregular celestial bodies, track the whole extent of the great solar province. To these might be added the whirlwinds of meteors, as it were disaggregated comets, which also circle round the Sun, and give origin to shooting stars, when they come into collision with the Earth.

Having now a general idea of our celestial family, and an appreciation of the potent focus that controls it, let us make direct acquaintance with the several members of which it is composed.

CHAPTER V

THE PLANETS

A.--MERCURY, VENUS, THE EARTH, MARS

And now we are in the Solar System, at the center, or, better, at the focus of which burns the immense and dazzling orb. We have appreciated the grandeur and potency of the solar globe, whose rays spread out in active waves that bear a fecundating illumination to the worlds that gravitate round him; we have appreciated the distance that separates the Sun from the Earth, the third of the planets retained within his domain, or at least I trust that the comparisons of the times required by certain moving objects to traverse this distance have enabled us to conceive it.

We said that the four planets nearest to the Sun are Mercury, at a distance of 57 million kilometers (35,000,000 miles); Venus, at 108 million (67,000,000 miles); the Earth, at 149 million (93,000,000 miles); and Mars at 226 million (140,000,000 miles). Let us begin our planetary journey with these four stations.

MERCURY

A little above the Sun one sometimes sees, now in the West, in the lingering shimmer of the twilight, now in the East, when the tender roseate dawn announces the advent of a clear day, a small star of the first magnitude which remains but a very short time above the horizon, and then plunges back into the flaming sun. This is Mercury, the agile and active messenger of Olympus, the god of eloquence, of medicine, of commerce, and of thieves. One only sees him furtively, from time to time, at the periods of his greatest elongations, either after the setting or before the rising of the radiant orb, when he presents the aspect of a somewhat reddish star.

This planet, like the others, shines only by the reflection of the Sun whose illumination he receives, and as he is in close juxtaposition with it, his light is bright enough, though his volume is inconsiderable. He is smaller than the Earth. His revolution round the Sun being accomplished in about three months, he passes rapidly, in a month and a half, from one side to the other of the orb of day, and is alternately a morning and an evening star. The ancients originally regarded it as two separate planets; but with attentive observation, they soon perceived its identity. In our somewhat foggy climates, it can only be discovered once or twice a year, and then only by looking for it according to the indications given in the astronomic almanacs.

Mercury courses round the Sun at a distance of 57,000,000 kilometers (35,000,000 miles), and accomplishes his revolution in 87 days, 23 hours, 15 minutes; i.e., 2 months, 27 days, 23 hours, or a little less than three of our months. If the conditions of life are the same there as here, the existence of the Mercurians must be four times as short as our own. A youth of twenty, awaking to the promise of the life he is just beginning in this world, is an octogenarian in Mercury. There the fair sex would indeed be justified in bewailing the transitory nature of life, and might regret the years that pass too quickly away. Perhaps, however, they are more philosophic than with us.

The orbit of Mercury, which of course is within that of the Earth, is not circular, but elliptical, and very eccentric, so elongated that at certain times of the year this planet is extremely remote from the solar focus, and receives only half as much heat and light as at the opposite period; and, in consequence, his distance from the Earth varies considerably.

This globe exhibits phases, discovered in the seventeenth century by Galileo, which recall those of the Moon. They are due to the motions of the planet round the Sun, and are invisible to the unaided eye, but with even a small instrument, one can follow the gradations and study Mercury under every aspect. Sometimes, again, he passes exactly in front of the Sun, and his disk is projected like a black point upon the luminous surface of the flaming orb. This occurred, notably, on May 10, 1891, and November 10, 1894; and the phenomenon will recur on November 12, 1907, and November 6, 1914.

Mercury is the least of all the worlds in our system (with the exception of the cosmic fragments that circulate between the orbit of Mars and that of Jupiter). His volume equals only 5/100 that of the Earth. His diameter, in comparison with that of our planet, is in the ratio of 373 to 1,000 (a little more than 1/3) and measures 4,750 kilometers (2,946 miles). His density is the highest of all the worlds in the great solar family, and exceeds that of our Earth by about 1/3; but weight there is less by almost 1/2.

Mercury is enveloped in a very dense, thick atmosphere, which doubtless sensibly tempers the solar heat, for the Sun exhibits to the Mercurians a luminous disk about seven times more extensive than that with which we are familiar on the Earth, and when Mercury is at perihelion (that is, nearest to the Sun), his inhabitants receive ten times more light and heat than we obtain at midsummer. In all probability, it would be impossible for us to set foot on this planet without being shattered by a sunstroke.

Yet we may well imagine that Nature's fecundity can have engendered beings there of an organization different from our own, adapted to an existence in the proximity of fire. What magnificent landscapes may there be adorned with the luxuriant vegetation that develops rapidly under an ardent and generous sun?

Observations of Mercury are taken under great difficulties, just because of the immediate proximity of the solar furnace; yet some have detected patches that might be seas. In any case, these observations are contradictory and uncertain.

Up to the present it has been impossible to determine the duration of the rotation. Some astronomers even think that the Sun's close proximity must have produced strong tides, that would, as it were, have immobilized the globe of Mercury, just as the Earth has immobilized the Moon, forcing it perpetually to present the same side to the Sun. From the point of view of habitation, this situation would be somewhat peculiar; perpetual day upon the illumined half, perpetual night upon the other hemisphere, and a fairly large zone of twilight between the two. Such a condition would indeed be different from the succession of terrestrial days and nights.

As seen from Mercury, the Earth we inhabit would shine out in the starry sky[9] as a magnificent orb of first magnitude, with the Moon alongside, a faithful little companion. They should form a fine double star, the Earth being a brilliant orb of first magnitude, and the Moon of third, a charming couple, and admired doubtless as an enchanted and privileged abode.

It is at midnight during the oppositions of the Earth with the Sun that our planet is the most beautiful and brilliant, as is Jupiter for ourselves. The constellations are the same, viewed from Mercury or from the Earth.

But is this little solar planet inhabited? We do not yet know. We can only reply: why not?

VENUS

When the sunset atmosphere is crimson with the glorious rays of the King of Orbs, and all Nature assumes the brooding veil of twilight, the most indifferent eyes are often attracted and captivated by the presence of a star that is almost dazzling, and illuminates with its white and limpid light the heavens darkened by the disappearance of the God of Day.

Hail, Venus, Queen of the Heavens! the "Shepherd's Star," gentle mother of the loves, goddess of beauty, eternally adored and cherished, sung and immortalized upon Earth, by poets and artists. Her splendid brilliancy attracted notice from earliest antiquity, and we find her, radiant and charming, in the works of the ancients, who erected altars to her and adorned their poetry with her grace and beauty. Homer calls her Callisto the Beautiful; Cicero names her Vesper, the evening star, and Lucifer, the star of the morning--for it was with this divinity as with Mercury. For a long while she was regarded as two separate planets, and it was only when it came to be observed that the evening and the morning star were always in periodic succession, that the identity of the orb was recognized.

Her radiant splendor created her mythological personality, just as the agility of Mercury created that of the messenger of the gods.

We do not see her aerial chariot in the Heavens drawn by a flight of doves with white and fluttering wings, but we follow the lustrous orb led on through space by solar attraction. And in the beautiful evenings when she is at her greatest distance from our Sun, the whole world admires this white and dazzling Venus reigning as sovereign over our twilight[10] for hours after sunset, and in addition to the savants who are practically occupied with astronomy, millions of eyes are raised to this celestial splendor, and for a moment millions of human beings feel some curiosity about the mysteries of the Infinite. The brutalities of daily life would fain petrify our dreams, but thought is not yet stifled to the point of checking all aspirations after eternal truth, and when we gaze at the starry sky it is hard not to ask ourselves the nature of those other worlds, and the place occupied by our own planet in the vast concert of sidereal harmony.

Fig. 37 gives some notion of the succession of these, and of the planet's variations in magnitude during its journey round the Sun. Imagine it to be rotating in a year of 224 days, 16 hours, 49 minutes, 8 seconds at a distance of 108 million kilometers (67,000,000 miles), the Earth being at 149 million kilometers (93,000,000 miles). Like Mercury, at certain periods it passes between the Sun and ourselves, and as its illuminated hemisphere is of course turned toward the orb of day, we at those times perceive only a sharp and very luminous crescent. At such periods Venus is entirely, so to say, against the Sun, and presents to us her greatest apparent dimension (Fig. 38). Sometimes, again, like Mercury, she passes immediately in front of the Sun, forming a perfectly round black spot; this happened on December 8, 1874, and December 6, 1882; and will recur on June 7, 2004, and June 5, 2012. These transits have been utilized in celestial geometry in measuring the distance of the Sun.

You will readily divine that the distance of Venus varies considerably according to her position in relation to the Earth: when she is between the Sun and ourselves she is nearest to our world; but it is just at those times that we see least of her surface, because she exhibits to us only a slender crescent. Terrestrial astronomers are accordingly very badly placed for the study of her physical constitution. The best observations can be made when she is situated to right or left of the Sun, and shows us about half her illuminated disk--during the day for choice, because at night there is too much irradiation from her dazzling light.

These phases were discovered by Galileo, in 1610. His observations were among the first that confirmed the veracity of the system of Copernicus, affording an evident example of the movement of the planets round the sun. They are often visible to the unaided eye with good sight, either at dusk, or through light clouds.

Venus, surrounded by a highly dense and rarefied atmosphere, which increases the difficulties of observing her surface, might be called the twin sister of the Earth, so similar are the dimensions of the two worlds. But, strange as it may seem to the many admirers, who are ready to hail in her an abode of joy and happiness, it is most probable that this planet, attractive as she is at a distance, would be a less desirable habitation than our floating island. In fact, the atmosphere of Venus is perpetually covered with cloud, so that the weather there must be always foggy. No definite geographical configuration can be discovered on her, despite the hopes of the eighteenth-century astronomers. We are not even sure that she rotates upon herself, so contradictory are the observations, and so hard is it to distinguish anything clearly upon her surface. A single night of observation suffices to show the rotation of Mars or of Jupiter; but the beautiful Evening Star remains obstinately veiled from our curiosity.

Several astronomers, and not the least considerable, think that the tides produced by the Sun upon her seas, or globe in its state of pristine fluidity, must have been strong enough to seize and fix her, as the Earth did for the Moon, thus obliging her to present always the same face to the Sun. Certain telescopic observations would even seem to confirm this theoretical deduction from the calculations of celestial mechanics.

The author ventures to disagree with this opinion, its apparent probability notwithstanding, because he has invariably received a contrary impression from all his telescopic observations. He has quite recently (spring of 1903) repeated these observations. Choosing a remarkably clear and perfectly calm atmosphere, he examined the splendid planet several times with great attention in the field of the telescope. The right or eastern border (reversed image) was dulled by the atmosphere of Venus; this is the line of separation between day and night. Beneath, at the extreme northern edge, he was attracted on each occasion by a small white patch, a little whiter than the rest of the surface of the planet, surrounded by a light-gray penumbra, giving the exact effect of a polar snow, very analogous to that observed at the poles of Mars. To the author this white spot on the boreal horn of Venus does not appear to be due to an effect of contrast, as has sometimes been supposed.

Now, if the globe of Venus has poles, it must turn upon itself.

Unfortunately it has proved impossible to distinguish any sign upon the disk, indicative of the direction and speed of its rotary movement, although these observations were made, with others, under excellent conditions.--Three o'clock in the afternoon, brilliant sun, sky clear blue, the planet but little removed from the meridian--at which time it is less dazzling than in the evening.

There is merely the impression; but it is so definite as to prevent the author from adopting the new hypothesis, in virtue of which the planet, as it gravitates round the Sun, presents always the same hemisphere.

If this hypothesis were a reality, Venus would certainly be a very peculiar world. Eternal day on the one side; eternal night on the other. Maximum light and heat at the center of the hemisphere perpetually turned to the Sun; maximum cold and center of night at the antipodes. This icy hemisphere would possibly be uninhabitable, but the resources of Nature are so prodigious, and the law of Life is so imperious, so persistent, under the most disadvantageous and deplorable terrestrial conditions, that it would be transcending our rights to declare an impossibility of existence, even in this eternal night. The currents of the atmosphere would no doubt suffice to set up perpetual changes of temperature between the two hemispheres, in comparison with which our trade-winds would be the lightest of breezes.

Yes, mystery still reigns upon this adjacent earth, and the most powerful instruments of the observatories of the whole world have been unable to solve it. All we know is that the diameter, surface, volume and mass of this planet, and its weight at the surface, do not differ sensibly from those that characterize our own globe: that this planet is sister to our own, and of the same order, hence probably formed of the same elements. We further know that, as seen from Venus (Fig. 39), the Earth on which we live is a magnificent star, a double orb more brilliant even than when viewed from Mercury. It is a dazzling orb of first magnitude, accompanied by its moon, a star of the second and a half magnitude.

And thus the worlds float on in space, distant symbols of hopes not realized on any one of them, all at different stages of their degree of evolution, representing an ever-growing progress in the sequence of the ages.

When we contemplate this radiant Venus, it is difficult, even if we can not form any definite idea as to her actual state as regards habitation, to assume that she must be a dreary desert, and not, on the contrary, to hail in her a celestial land, differing more or less from our own dwelling-place, travailing with her sisters in the accomplishment of the general plan of Nature.

Such are the characteristic features of our celestial neighbor. In quitting her, we reach the Earth, which comes immediately next her in order of distance,

149 million kilometers (93,000,000 miles) from the Sun, but as we shall devote an entire chapter to our own planet, we will not halt at this point, but cross in one step the distance that separates Mars from Venus.

Let us only remark in passing, that our planet is the largest of the four spheres adjacent to the Sun. Here are their comparative diameters:

The Earth = 1. In Kilometers. In Miles. Mercury 0.373 4,750 2,946 Venus 0.999 12,730 7,894 Earth 1.000 12,742 7,926 Mars 0.528 6,728 4,172

It will be seen that Venus is almost identical with the Earth.

MARS

Two hundred and twenty-six millions of kilometers (140,000,000 miles) from the Sun is the planet Mars, gravitating in an orbit exterior to that which the Earth takes annually round the same center.

Unfortunate Mars! What evil fairy presided at his birth? From antiquity, all curses seem to have fallen upon him. He is the god of war and of carnage, the protector of armies, the inspirer of hatred among the peoples, it is he who pours out the blood of Humanity in international hecatombs. Here, again, as in the case of Mercury and Venus, the appearance has originated the idea. Mars, in fact, burns like a drop of blood in the depths of the firmament, and it is this ruddy color that inspired its name and attributes, just as the dazzling whiteness of Venus made her the goddess of love and beauty. Why, indeed, should the origins of mythology be sought elsewhere than in astronomy?

While Humanity was attributing to the presumptive influence of Mars the defects inherent in its own terrestrial nature, this world, unwitting of our sorrows, pursued the celestial path marked out for it in space by destiny.

This planet is, as we have said, the first encountered after the Earth. Its orbit is very elongated, very eccentric. Mars accomplishes it in a period of 1 year, 321 days, 22 hours, i.e., 1 year, 10 months, 21 days, or 687 days. The velocity of its transit is 23 kilometers (14.5 miles) per second; that of the Earth is 30 (19 miles). Our planet, traveling through space at an average distance of 149 million kilometers (93,000,000 miles) from the central focus, is separated from

Mars by an average distance of 76 million kilometers (47,000,000 miles); but as its orbit is equally elliptic and elongated it follows that at certain epochs the two planets approach one another by something less than 60 million kilometers (37,000,000 miles). These are the periods selected for making the best observations upon our neighbor of the ruddy rays. The oppositions of Mars arrive about every twenty-six months, but the periods of its greatest proximity, when this planet approaches to within 56 million kilometers (34,700,000 miles) of the Earth, occur only every fifteen years.

Mars is then passing perihelion, while our world is at aphelion (or greatest distance from the Sun). At such epochs this globe presents to us an apparent diameter 63 times smaller than that of the Moon, i.e., a telescope that magnifies 63 times would show him to us of the same magnitude as our satellite viewed with the unaided eye, and an instrument that magnified 630 times would show him ten times larger in diameter.

In dimensions he differs considerably from our world, being almost half the size of the Earth. In diameter he measures only 6,728 kilometers (4,172 miles), and his circumference is 21,125 kilometers (13,000 miles). His surface is only 29/100 of the terrestrial surface, and his volume only 15/100 of our own.

This difference in volume causes Mars to be an earth in miniature. When we study his aspects, his geography, his meteorology, we seem to see in space a reduction of our own abode, with certain dissimilarities that excite our curiosity, and make him even more interesting to us.

The Martian world weighs nine times and a half less than our own. If we represent the weight of the Earth by 1,000, that of Mars would be represented by 105. His density is much less than our own; it is only 7/10 that of the Earth. A man weighing 70 kilograms, transported to the adjacent globe, would weigh only 26 kilograms.

The earliest telescopic observations revealed the existence of more or less accentuated markings upon the surface of Mars. The progress of optics, admitting of greater magnifications, exhibited the form of these patches more clearly, while the study of their motions enabled the astronomers to determine with remarkable precision the diurnal rotation of this planet. It occurs in 24 hours, 37 minutes, 23.65 seconds. Day and night are accordingly a little longer on Mars than on the Earth, but the difference is obviously inconsiderable. The year of Mars consists of 668 Martian days. The inclination of the axis of rotation of this globe upon the plane of its orbit is much the same as our own. In consequence, its seasons are analogous to ours in intensity, while twice the length, the Martian year being almost equal to two of our years. The intensity of the seasons is indeed more accentuated than upon the Earth, since the orbit of Mars is very elongated. But there, as here, are three quite distinct zones: the torrid, the temperate, and the glacial.

By means of the telescope we can follow the variations of the Martian seasons, especially in what concerns the polar snows, which regularly aggregate during the winter, and melt no less regularly during the heat of the summer. These snows are very easily observed, and stand out clearly with dazzling whiteness. The reader can judge of them by the accompanying figure, which sums up the author's observations during one of the recent oppositions of Mars (1900-1901). The size of the polar cap diminished from 4,680 kilometers to 840. The solstice of the Martian summer was on April 11th. The snows were still melting on July 6th. Sometimes they disappear almost entirely during the Martian month that corresponds to our month of August, as never happens with our polar ice. Hence, though this planet is farther away from the Sun than ourselves, it does not appear to be colder, or, at any rate, it is certain that the polar snows are much less thick.

On the other hand, there are hardly ever clouds on Mars; the Martian atmosphere is almost always limpid, and one can say that fine weather is the chronic state of the planet. At times, light fogs or a little vapor will appear in certain regions, but they are soon dissipated, and the sky clears up again.

Since the invention of the telescope, a considerable number of drawings have been made, depicting Mars under every aspect, and the agreement between these numerous observations gives us a sufficient acquaintance with the planet to admit of our indicating the characteristic features of its geography, and of drawing out areographic maps (Ares, Mars). Its appearance can be judged of from the two drawings here reproduced, as made (February, 1901) at the Observatory of Juvisy, and from the general chart drawn from the total sum of observations (Figs. 41, 42 and 43).

It will be seen at the first glance that the geography of Mars is very different from that of our own globe: while three-quarters of the Earth are covered with the liquid element, Mars seems to be more evenly divided, and must indeed have rather more land than water. We find no immense oceans surrounding the continents, and separating them like islands; on the contrary, the seas are reduced to long gulfs compressed between the shores, like the Mediterranean for example, nor is it even certain that these gray spots do all represent true seas. It has been agreed to term sea the parts that are lightly tinged with green, and to give the name of continent to the spots colored yellow. That is the hue of the Martian soil, due either to the soil itself, which would resemble that of the Sahara, or, to take a less arid region, that seen on the line between Marseilles and Nice, in the vicinity of the Esterels; or perhaps to some peculiar vegetation. During ascents in a balloon, the author has often remarked that the hue of the ripe corn, with the Sun shining on it, is precisely that presented to us by the continents of Mars in the best hours for observation.

As to the "seas," it is pretty certain that there must be water, or some kind of liquid, deriving above all from the melting of the polar snows in spring and summer; but it may possibly be in conjunction with some vegetation, aquatic plants, or perhaps vast meadows, which appear to us from here to be the more considerable in proportion as the water that nourishes them has been more abundant.

Mars, like our globe, is surrounded with a protective atmosphere, which retains the rays of the Sun, and must preserve a medium temperature favorable to the conservation of life upon the surface of the planet. But the circulation of the water, so important to terrestrial life, whether animal or vegetable, which is effected upon our planet by the evaporation of the seas, clouds, winds, rains, wells, rivers and streams, comes about quite differently on Mars; for, as was remarked above, it is rarely that any clouds are observed there. Instead of being vertical, as here, this circulation is horizontal: the water coming from the source of the polar snows finds its way into the canals and seas, and returns to be condensed at the poles by a light drift of invisible vapors directed from the equator to the poles. There is never any rain.

We have spoken of canals. One of the great puzzles of the Martian world is incontestably the appearance of straight lines that furrow its surface in all directions, and seem to connect the seas. M. Schiaparelli, the distinguished Director of the Observatory of Milan, who discovered them in 1877, called them canals, without, however, postulating anything as to their real nature. Are they indeed canals? These straight lines, measuring sometimes 600 kilometers (372 miles) in length, and more than 100 kilometers (62 miles) in breadth, have much the same hue as the seas on which they open. For a quarter of a century they have been surveyed by the greater number of our observers. But it must be confessed that, even with the best instruments, we only approach Mars at a distance of 60,000 kilometers (37,200 miles), which is still a little far off, and we may be sure that we do not distinguish the true details of the surface.[11] These details at the limits of visibility produce the appearance of canals to our eyes. They may possibly be lines of lakes, or oases. The future will no doubt clear up this mystery for us.

As to the inhabitants of Mars, this world is in a situation as favorable as our Earth for habitation, and it would be difficult to discover any reason for perpetual sterility there. It appears to us, on the contrary, by its rapid and frequent variations of aspect, to be a very living world. Its atmosphere, which is always clear, has not the density of our own, and resembles that of the highest mountains. The conditions of existence there vary from ours, and appear to be more delicate, more ethereal.

There as here, day succeeds to night, spring softens the rigors of winter; the seasons unfold, less disparate than our own, of which we have such frequent reason to complain. The sky is perpetually clear. There are never tempests, hurricanes, nor cyclones, the wind never gets up any force there, on account of the rarity of the atmosphere, and the low intensity of weight.

Differing from ours, this world may well be a more congenial habitation. It is more ancient than the Earth, smaller, less massive. It has run more quickly through the phases of its evolution. Its astral life is more advanced, and its Humanity should be superior to our own, just as our successors a million years hence, for example, will be less coarse and barbarous than we are at present: the law of progress governs all the worlds, and, moreover, the physical constitution of the planet Mars is less dense than our own.

There is no need to despair of entering some day into communication with these unknown beings. The luminous points that have been observed are no signals, but high summits or light clouds illuminated by the rising or setting sun. But the idea of communication with them in the future is no more audacious and no less scientific than the invention of spectral analysis, X-rays, or wireless telegraphy.

We may suppose that the study of astronomy is further advanced in Mars than on the Earth, because humanity itself has advanced further, and because the starry sky is far finer there, far easier to study, owing to the limpidity of its pure, clear atmosphere.

Two small moons (hardly larger than the city of Paris) revolve rapidly round Mars; they are called Phobos and Deimos. The former, at a distance of 6,000 kilometers (3,730 miles) from the surface, accomplishes its revolution rapidly, in seven hours, thirty-nine minutes, and thus makes the entire circle of the Heavens three times a day. The second gravitates at 20,000 kilometers (12,400 miles), and turns round its center of attraction in thirty hours and eighteen minutes. These two satellites were discovered by Mr. Hall, at the University of Washington, in the month of August, 1877.

* * * * *

Among the finest and most interesting of the celestial phenomena admired by the Martians, at certain epochs of the year,--now at night when the Sun has plunged into his fiery bed, now in the morning, a little before the aurora,--is a magnificent star of first magnitude, never far removed from the orb of day, which presents to them the same aspects as does Venus to ourselves. This splendid orb, which has doubtless received the most flattering names from those who contemplate it, this radiant star of a beautiful greenish blue, courses in space accompanied by a little satellite, sparkling like some splendid diamond, after sunset, in the clear sky of Mars. This superb orb is the Earth, and the little star accompanying it is the Moon.

Yes, to the Martians our Earth is a star of the morning and evening; doubtless they have determined her phases. Many a vow, and many a hope must have been wafted toward her, more than one broken heart must have permitted its unrealized dreams to wander forth to our planet as to an abode of happiness where all who have suffered in their native world might find a haven. But our planet, alas! is not as perfect as they imagine.

We must not dally upon Mars, but hasten our celestial excursion toward

Jupiter.

CHAPTER VI

THE PLANETS

B.--JUPITER, SATURN, URANUS, NEPTUNE.

Before we attack the giant world of our system, we must halt for a few moments upon the minor planets which circulate between the orbit of Mars and that of Jupiter. These minute asters, little worlds, the largest of which measures scarcely more than 100 kilometers (62 miles) in diameter, are fragments of cosmic matter that once belonged to a vast ring, formed at the time when the solar system was only an immense nebula; and which, instead of condensing into a single globe coursing between Mars and Jupiter, split up into a considerable quantity of particles constituting at the present time the curious and highly interesting Republic of the Asteroids.

These lilliputian worlds at first received the names of the more celebrated of the minor mythological divinities--Ceres, Pallas, Juno, Vesta, etc., but as they rapidly increased in number, it was found necessary to call them by modern, terrestrial names, and more than one daughter of Eve, the Egeria of some astronomer, now has her name inscribed in the Heavens. The first minor planet was discovered on the first day of the nineteenth century, January 1, 1801, by Piazzi, astronomer at Palermo. While he was observing the small stars in the constellation of the Bull beneath the clear Sicilian skies, this famous astronomer noticed one that he had never seen before.

The next night, directing his telescope to the same part of the Heavens, he perceived that the fair unknown had moved her station, and the observations of the following days left him no doubt as to the nature of the visitor: she was a planet, a wandering star among the constellations, revolving round the Sun. This newcomer was registered under the name of Ceres.

Since that epoch several hundreds of them have been discovered, occupying a zone that extends over a space of more than 400 million kilometers (249,000,000 miles). These celestial globules are invisible to the naked eye, but no year passes without new and numerous recruits being added to the already important catalogue of these minute asters by the patient observers of the Heavens. To-day, they are most frequently discovered by the photographic method of following the displacement of the tiny moving points upon an exposed sensitive plate.

JUPITER

And now let us bow respectfully before Jupiter, the giant of the worlds. This glorious planet is indeed King of the Solar System.

While Mercury measures only 4,750 kilometers (2,946 miles) in diameter, and Mars 6,728 kilometers (4,172), Jupiter is no less than 140,920 kilometers (87,400 miles) in breadth; that is to say, eleven times larger than the Earth. He is 442,500 kilometers (274,357 miles) in circumference.

In volume he is equivalent to 1,279 terrestrial globes; hence he is only a million times smaller than the Sun. The previously described planets of our system, Mercury, Venus, the Earth, and Mars combined, would form only an insignificant mass in comparison with this colossus. A hundred and twenty-six Earths joined into one group would present a surface whose extent would still not be quite as vast as the superficies of this titanic world. This immense globe weighs 310 times more than that which we inhabit. Its density is only the quarter of our own; but weight is twice and a half times as great there as here. The constituents of things and beings are thus composed of materials lighter than those upon the Earth; but, as the planet exerts a force of attraction twice and a half times as powerful, they are in reality heavier and weigh more. A graceful maiden weighing fifty kilograms would if transported to Jupiter immediately be included in the imposing society of the "Hundred Kilos."

Jupiter rotates upon himself with prodigious rapidity. He accomplishes his diurnal revolution in less than ten hours! There the day lasts half as long as here, and while we reckoned fifteen days upon our calendar, the Jovian would count thirty-six. As Jupiter's year equals nearly twelve of ours, the almanac of that planet would contain 10,455 days! Obviously, our pretty little pocket calendars would never serve to enumerate all the dates in this vast world.

This splendid globe courses in space at a distance of 775,000,000 kilometers (480,500,000 miles) from the Sun. Hence it is five times (5.2) as remote from the orb of day as our Earth, and its orbit is five times vaster than our own. At that distance the Sun subtends a diameter five times smaller than that which we see, and its surface is twenty-seven times less extensive; accordingly this planetary abode receives on an average twenty-seven times less light and heat than we obtain.

In the telescope Jupiter presents an aspect analogous to that likely to be exhibited by a world covered with clouds, and enveloped in dense vapors (Fig. 45).

It is, in fact, the seat of formidable perturbations, of strange revolutions by which it is perpetually convulsed, for although of more ancient formation than the Earth, this celestial giant has not yet arrived at the stable condition of our dwelling-place. Owing to its considerable volume, this globe has probably preserved its original heat, revolving in space as an obscure Sun, but perhaps still burning. In it we see what our own planet must have been in its primordial epoch, in the pristine times of terrestrial genesis.

Since its orbital revolution occupies nearly twelve years, Jupiter comes back into opposition with the Sun every 399 days, i.e., 1 year, 34 days, that is with one month and four days' delay each year. At these periods it is located at the extremity of a straight line which, passing by the Earth, is prolonged to the Sun. These are the epochs to be selected for observation. It shines then, all night, like some dazzling star of the first magnitude, of excessive whiteness: nor can it be confounded either with Venus, more luminous still (for she is never visible at midnight, in the full South, but is South-west in the evening, or South-east in the morning), nor with Mars, whose fires are ruddy.

In the telescope, the immense planet presents a superb disk that an enlargement of forty times shows us to be the same size to all appearance as that of the Moon seen with the unaided eye. Its shape is not absolutely spherical, but spheroid--that is, flattened at the poles. The flattening is 1/17.

We know that the Earth's axis dips a certain quantity on the plane of her orbit, and that it is this inclination that produces the seasons. Now it is not the same for Jupiter. His axis of rotation remains almost vertical throughout the course of his year, and results in the complete absence of climates and seasons. There is neither glacial zone, nor tropic zone; the position of Jupiter is eternally that of the Earth at the season of the equinox, and the vast world enjoys, as it were, perpetual spring. It knows neither the hoar-frost nor the snows of winter. The heat received from the Sun diminishes gradually from the equator to the poles without abrupt transitions, and the duration of day and night is equal there throughout the entire year, under every latitude. A privileged world, indeed!

It is surrounded by a very dense, thick atmosphere, which undergoes more extensive variations than could be produced by the Sun at such a distance. Spectral analysis detects a large amount of water-vapor, showing that this planet still possesses a very considerable quantity of intrinsic heat.

Most conspicuous upon this globe are the larger or smaller bands or markings (gray and white, sometimes tinted yellow, or of a maroon or chocolate hue) by which its surface is streaked, particularly in the vicinity of the equator. These different belts vary, and are constantly modified, either in form or color. Sometimes, they are irregular, and cut up; at others they are interspersed with more or less brilliant patches. These patches are not affixed to the surface of the globe, like the seas and continents of the Earth; nor do they circulate round the planet like the satellites, in more or less elongated and regular revolutions, but are relatively mobile, like our clouds in the atmosphere, while observation of their motion does not give the exact period of the rotation of Jupiter. Some only appear upon the agitated disk to vanish very quickly; others subsist for a considerable period.

One has been observed for over a quarter of a century, and appears to be almost immobile upon this colossal globe. This spot, which was red at its first appearance, is now pale and ghostly. It is oval (vide Fig. 45) and measures 42,000 kilometers (26,040 miles) in length by 15,000 kilometers (9,300 miles) in width. Hence it is about four times as long as the diameter of our Earth; that is, relatively to the size of Jupiter, as are the dimensions of Australia in proportion to our globe. The discussion of a larger number of observations leads us to see in it a sort of continent in the making, a scoria recently ejected from the mobile and still liquid and heated surface of the giant Jupiter. The patch, however, oscillates perceptibly, and appears to be a floating island.

We must add that this vast world, like the Sun, does not rotate all in one period. Eight different currents can be perceived upon its surface. The most rapid is that of the equatorial zone, which accomplishes its revolution in 9 hours, 50 minutes, 29 seconds. A point situated on the equator is therefore carried forward at a speed of 12,500 meters (7 miles) per second, and it is this giddy velocity of Jupiter that has produced the flattening of the poles. From the equator to the poles, the swiftness of the currents diminishes irregularly, and the difference amounts to about five minutes between the movement of the equatorial stream, and that of the northern and southern currents. But what is more curious still is that the velocity of one and the same stream is subject to certain fluctuations; thus, in the last quarter of a century, the speed of the equatorial current has progressively diminished. In 1879, the velocity was 9 hours, 49 minutes, 59 seconds, and now it is, as we have already seen, 9 hours, 50 minutes, 29 seconds, which represents a substantial reduction. The rotation of the red patch, at 25 degrees of the southern latitude, is effected in 9 hours, 55 minutes, 40 seconds.

We are confronted with a strange and mysterious world. It is the world of the future.

This giant gravitates in space accompanied by a suite of five satellites. These are:

Names. Distance from surface of Jupiter. Time of revolution. Kilometers. Miles. Days. Hours. 5. 200,000 124,000 11 1. Io 430,000 266,000 1 18 2. Europa 682,000 422,840 3 13 3. Ganymede 1,088,000 674,560 7 4 4. Callisto 1,914,000 1,186,680 16 16

The four principal satellites of Jupiter were discovered at the same time, on the same evenings (January 7 and 8, 1610), by the two astronomers who were pointing their telescopes at Jupiter: Galileo in Italy, and Simon Marius in Germany.

On September 9, 1892, Mr. Barnard, astronomer of the Lick Observatory, California, discovered a new satellite, extremely minute, and very near the enormous planet. It has so far received no name, and is known as the fifth, although the four principal are numbered in the order of their distances.

The four classical satellites are visible in the smallest instruments (Fig. 46): the third is the most voluminous.

Such is the splendid system of the mighty Jupiter. Once, doubtless, this fine planet illuminated the troop of worlds that derived their treasure of vitality from him with his intrinsic light: to-day, however, these moons in their turn shed upon the extinct central globe the pale soft light which they receive from our solar focus, illuminating the brief Jovian nights (which last less than five hours, on account of the twilight) with their variable brilliancy.

At the distance of the first satellite, Jupiter exhibits a disk fourteen hundred times vaster than that of the Full Moon! What a dazzling spectacle, what a fairy scene must the enormous star afford to the inhabitants of that tiny world! And what a shabby figure must our Earth and Moon present in the face of such a body, a real miniature of the great solar system!

Our ancestors were well inspired when they attributed the sovereignty of Olympus to this majestic planet. His brilliancy corresponds with his real grandeur. His dominion in the midnight Heavens is unique. Here again, as for Venus, Mars, and Mercury, astronomy has created the legend of the fables of mythology.

Let us repeat in conclusion that our Earth becomes practically invisible for the inhabitants of the other worlds beyond the distance of Jupiter.

SATURN

Turn back now for a moment to the plan of the Solar System.

We had to cross 775 million kilometers (480,000,000 miles) when we left the Sun, in order to reach the immense orb of Jupiter, which courses in space at 626 million kilometers (388,000,000 miles) from the terrestrial orbit. From Jupiter we had to traverse a distance of 646 million kilometers (400,000,000 miles) in order to reach the marvelous system of Saturn, where our eyes and thoughts must next alight.

Son of Uranus and Vesta, Saturn was the God of Time and Fate. He is generally represented as an aged man bearing a scythe. His mythological character is only the expression of his celestial aspect, as we have seen for the brilliant Jupiter, for the pale Venus, the ruddy Mars, and the agile Mercury. The revolution of Saturn is the slowest of any among the planets known to the ancients. It takes almost thirty years for its accomplishment, and at that distance the Saturnian world, though it still shines with the brilliancy of a star of the first magnitude, exhibits to our eyes a pale and leaden hue. Here is, indeed, the god of Time, with slow and almost funereal gait.

Poor Saturn won no favor with the poets and astrologers. He bore the horrid reputation of being the inexhaustible source of misfortune and evil fates,-- whereof he is wholly innocent, troubling himself not at all with our world nor its inhabitants.

This world travels in the vastness of the Heavens at a distance of 1,421 million kilometers (881,000,000 miles) from the Sun. Hence it is ten times farther from the orb of day than the Earth, though still illuminated and governed by the Sun-God. Its gigantic orbit is ten times larger than our own.

Its revolution round the Sun is accomplished in 10,759 days, i.e., 29 years, 167 days, and as this strange planet rotates upon itself with great rapidity in 10 hours, 15 minutes, its year comprises no less than 25,217 days. What a calendar! The Saturnians must needs have a prodigious memory not to get hopelessly involved in this interminable number of days. A curious world, where each year stands for almost thirty of our own, and where the day is more than half as short again as ours. But we shall presently find other and more extraordinary differences on this planet.

In the first place it is nearly nine and a half times larger than our world. It is a globe, not spherical, but spheroidal, and the flattening of its poles, which is one-tenth, exceeds that of all the other planets, even Jupiter. It follows that its equatorial diameter is 112,500 kilometers (69,750 miles), while its polar diameter measures only 110,000 kilometers (68,200).

In volume, Saturn is 719 times larger than the Earth, but its density is only 128/1000 of our own; i.e., the materials of which it is composed are much less heavy, so that it weighs only 92 times more than our Earth. Its surface is 85 times vaster than that of the Earth, no insignificant proportion.

The dipping of Saturn's axis of rotation is much the same as our own. Hence we conclude that the seasons of this planet are analogous to ours in relative intensity. Only upon this far-off world each season lasts for seven years. At the distance at which it gravitates in space, the heat and light which it receives from the Sun are 90 times less active than such as reach our selves; but it apparently possesses an atmosphere of great density, which may be constituted so that the heat is preserved, and the planet maintained in a calorific condition but little inferior to our own.

In the telescope, the disk of Saturn exhibits large belts that recall those of Jupiter, though they are broader and less accentuated (Fig. 47). There are doubtless zones of clouds or rapid currents circulating in the atmosphere. Spots are also visible whose displacement assists in calculating the diurnal motions of this globe.

The most extraordinary characteristic of this strange world is, however, the existence of a vast ring, which is almost flat and very large, and entirely envelops the body of the planet. It is suspended in the Saturnian sky, like a gigantic triumphal arch, at a height of some 20,000 kilometers (12,400 miles) above the equator. This splendid arch is circular, like an immense crown illuminated by the Sun. From here we only see it obliquely, and it appears to us elliptical; a part of the ring seems to pass in front of Saturn, and its shadow is visible on the planet, while the opposite part passes behind.

This ring, which measures 284,000 kilometers (176,080 miles) in diameter, and less than 100 kilometers (62 miles) in breadth, is divided into three distinct zones: the exterior is less luminous than the center, which is always brighter than the planet itself; the interior is very dark, and spreads out like a dusky and faintly transparent veil, through which Saturn can be distinguished.

What is the nature of these vast concentric circles that surround the planet with a luminous halo? They are composed of an innumerable number of particles, of a quantity of cosmic fragments, which are swept off in a rapid revolution, and gravitate round the planet at variable speed and distance. The nearer particles must accomplish their revolution in 5 hours, 50 minutes, and the most distant in about 12 hours, 5 minutes, to prevent them from being merged in the surface of Saturn: their own centrifugal force sustains them in space.

[Illustration: FIG. 48. Varying perspective of Saturn's Rings, as seen from the Earth.]

With a good glass the effect of these rings is most striking, and one can not refrain from emotion on contemplating this marvel, whereby one of the brothers of our terrestrial country is crowned with a golden diadem. Its aspects vary with its perspective relative to the Earth, as may be seen from the subjoined figure (Fig. 48).

We must not quit the Saturnian province without mentioning the eight satellites that form his splendid suite:

Names. Distance from the planet. Time of revolution. Kilometers. Miles. Days. Hours. Minutes. 1. Mimas 207,000 128,340 22 37 2. Enceladus 257,600 159,712 1 8 53 3. Tethys 328,800 203,856 1 21 18 4. Dione 421,200 261,144 2 17 41 5. Rhea 588,400 364,808 4 12 25 6. Titan 1,364,000 845,680 15 22 41 7. Hyperion 1,650,000 1,023,000 21 6 39 8. Japhet 3,964,000 2,457,680 79 7 54

Here is a marvelous system, with, what is more, eight different kinds of months for the inhabitants of Saturn; eight moons with constantly varying phases juggling above the rings!

Now we shall cross at a bound the 1,400 million kilometers (868,000,000 miles) that separate us from the last station but one of the immense solar system.

URANUS

On March 13, 1781, William Herschel, a Hanoverian astronomer who had emigrated to England, having abandoned the study of music to devote himself to the sublime science of the Heavens, was observing the vast fields with their constellations of golden stars, when he perceived a luminous point that appeared to him to exceed that of the other celestial luminaries in diameter. He replaced the magnification of his telescope by more powerful eye-pieces, and found that the apparent diameter of the orb increased proportionately with the amplification of the power, which does not happen in the case of stars at infinite distance. His observations on the following evenings enabled him to note the slow and imperceptible movement of this star upon the celestial sphere, and left him in no further doubt: there was no star, but some much nearer orb, in all probability a comet, for the great astronomer dared not predict the discovery of a new planet. And it was thus, under the name of cometary orb, that the seventh child of the Sun was announced. The astronomers sought to determine the motions of the new arrival, to discover for it an elliptical orbit such as most comets have. But their efforts were vain, and after several months' study the conclusion was reached that here was a new planet, throwing back the limits of the solar system to a point far beyond that of the Saturnian frontier, as admitted from antiquity.

This new world received the name of Uranus, father of Saturn, his nearest neighbor in the solar empire. Uranus shines in the firmament as a small star of sixth magnitude, invisible to the unaided eye for normal sight, at a distance of 2,831,000,000 kilometers (1,755,000,000 miles) from the Sun. Smaller than Jupiter and Saturn, this planet is yet larger than Mercury, Venus, Mars, and the Earth together, thus presenting proportions that claim our respect and admiration.

His diameter may be taken at about 55,000 kilometers (34,200 miles), that is, rather more than four times the breadth of the terrestrial diameter. Sixty-nine times more voluminous than the Earth, and seventeen times more extensive in surface, this new world is much less than our own in density. The matter of which it is composed is nearly five times lighter than that of our globe.

Spectral analysis shows that this distant planet is surrounded with an atmosphere very different from that which we breathe, enclosing gases that do not exist in ours.

The Uranian globe courses over the fields of infinity in a vast orbit seventeen times larger than our own, and its revolution lasts 36,688 days, i.e., 84 years, 8 days. It travels slowly and sadly under the pale and languishing rays of the Sun, which sends it nearly three hundred times less of light and heat than we receive. At this distance the solar disk would present a diameter seventeen times smaller than that which we admire, and a surface three hundred times less vast. A dull world indeed! And what an interminable year! The idle people who are in the habit of being bored must find time even longer upon Uranus than upon our little Earth, where the days pass so rapidly. And if matters are arranged there as here, a babe of a year old, beginning to babble in its nurse's arms, would already have lived as long as an old man of eighty-four in this world.

But what most seriously complicates the Calendar of the Uranians is the fact that the four moons which accompany the planet accomplish their revolution in four different kinds of months, in two, four, eight, and thirteen days, as is shown in the following table:

Distance from the planet. Time of revolution. Kilometers. Miles. Days. Hours. Minutes.

1. Ariel 196,000 121,520 2 12 29 2. Umbriel 276,000 171,120 4 3 27 3. Titania 450,000 279,000 8 16 56 4. Oberon 600,000 372,000 13 11 7

The most curious fact is that these satellites do not rotate like those of the other planets. While the moons of the Earth, Mars, Jupiter, and Saturn accomplish their revolution from east to west, the satellites of Uranus rotate in a plane almost perpendicular to the ecliptic, and it is doubtless the same for the rotation of the planet.

If we had to quit the Earth, and fixate ourselves upon another world, we should prefer Mars to Uranus, where everything must be so different from terrestrial arrangements? But who knows? Perhaps, after all, this planet might afford us some agreeable surprises. Il ne faut jurer de rien.

NEPTUNE

And here we reach the frontier of the Solar System, as actually known to us. In landing on the world of Neptune, which circles through the Heavens in eternal twilight at a distance of more than four milliard kilometers (2,480,000,000 miles) from the common center of attraction of the planetary orbs, we once again admire the prodigies of science.

Uranus was discovered with the telescope, Neptune by calculation. In addition to the solar influence, the worlds exert a mutual attraction upon each other that slightly deranges the harmony ordered by the Sun. The stronger act upon the weaker, and the colossal Jupiter alone causes many of the perturbations in our great solar family. Now during regular observations of the position of Uranus in space, some inexplicable irregularities were soon perceived. The astronomers having full faith in the universality of the law of attraction, could not do otherwise than attribute these irregularities to the influence of some unknown planet situated even farther off. But at what distance?

A very simple proportion, known as Bode's law, has been observed, which indicates approximately the relative distances of the planets from the Sun. It is as follows: Starting from 0, write the number 3, and double successively,

0 3 6 12 24 48 96 192 384.

Then, add the number 4 to each of the preceding figures, which gives the following series:

4 7 10 16 28 52 100 196 388.

Now it is a very curious fact that if the distance between the Earth and the Sun be represented by 10, the figure 4 represents the orbit of Mercury, 7 that of Venus, 16 of Mars; the figure 28 stands for the medium distance of the minor planets; the distances of Jupiter, Saturn, and Uranus agree with 52, 100, and 196.

The immortal French mathematician Le Verrier, who pursued the solution of the Uranian problem, supposed naturally that the disturbing planet must be at the distance of 388, and made his calculations accordingly. Its direction in the Heavens was indicated by the form of the disturbances; the orbit of Uranus bulging, as it were, on the side of the disturbing factor.

On August 31, 1846, Le Verrier announced the position of the ultra-Uranian planet, and on September 23d following, a German astronomer, Galle, at the Observatory of Berlin, who had just received this intelligence, pointed his telescope toward the quarter of the Heavens designated, and, in fact, attested the presence of the new orb. Without quitting his study table, Le Verrier, by the sole use of mathematics, had detected, and, as it were, touched at pen's point the mysterious stranger.

Only, it is proved by observation and calculation that it is less remote than was expected from the preceding law, for it gravitates at a distance of 300, given that from the Earth to the Sun as 10.

This planet was called Neptune, god of the seas, son of Saturn, brother of Jupiter. The name is well chosen, since the King of the Ocean lives in darkness in the depths of the sea, and Le Verrier's orb is also plunged in the semi-obscurity of the depths of the celestial element. But it was primarily selected to do justice to an English astronomer, Adams, who had simultaneously made the same calculations as Le Verrier, and obtained the same results--without publishing them. His work remained in the records of the Greenwich Observatory.

The English command the seas, and wherever they dip their finger into the water and find it salt, they feel themselves "at home," and know that "Neptune's trident is the scepter of the world," hence this complimentary nomenclature.

Neptune is separated by a distance of four milliards, four hundred million kilometers from the solar center.

At such a distance, thirty times greater than that which exists between the Sun and our world, Neptune receives nine hundred times less light and heat than ourselves; i.e., Spitzbergen and the polar regions of our globe are furnaces compared with what must be the Neptunian temperature. Absolutely invisible to the unaided eye, this world presents in the telescope the aspect of a star of the eighth magnitude. With powerful magnifications it is possible to measure its disk, which appears to be slightly tinged with blue. Its diameter is four times larger than our own, and measures about 48,000 kilometers (29,900 miles), its surface is sixteen times vaster than that of the Earth, and to attain its volume we should have to put together fifty-five globes similar to our own. Weight at its surface must be about the same as here, but its medium density is only 1/3 that of the Earth.

It gravitates slowly, dragging itself along an orbit thirty times vaster than that of our globe, and its revolution takes 164 years, 281 days, i.e., 164 years, 9 months. A single year of Neptune thus covers several generations of terrestrial life. Existence must, indeed, be strange in that tortoise-footed world!

While in their rotation period, Mercury accomplishes 47 kilometers (29-3/8 miles) per second, and the Earth 29-1/2 (18-1/8 miles), Neptune rolls along his immense orbit at a rate of only 5-1/2 kilometers (about 3-1/4 miles) per second.

The vast distance that separates us prevents our distinguishing any details of his surface, but spectral analysis reveals the presence of an absorbent atmosphere in which are gases unknown to the air of our planet, and of which the chemical composition resembles that of the atmosphere of Uranus.

One satellite has been discovered for Neptune. It has a considerable inclination, and rotates from east to west.

* * * *

And here we have reached the goal of our interplanetary journey. After visiting the vast provinces of the solar republic, we feel yet greater admiration and gratitude toward the luminary that governs, warms, and illuminates the worlds of his system.

In conclusion, let us again insist that the Earth,--a splendid orb as viewed from Mercury, Venus, and Mars,--begins to disappear from Jupiter, where she becomes no more than a tiny spark oscillating from side to side of the Sun, and occasionally passing in front of him as a small black dot. From Saturn the visibility of our planet is even more reduced. As to Uranus and Neptune, we are invisible there, at least to eyes constructed like our own. We do not possess in the Universe the importance with which we would endow ourselves.

Neptune up to the present guards the portals of our celestial system; we will leave him to watch over the distant frontier; but before returning to the Earth, we must glance at certain eccentric orbs, at the mad, capricious comets, which imprint their airy flight upon the realms of space.

CHAPTER VII

THE COMETS

SHOOTING STARS, BOLIDES, URANOLITHS OR METEORIC STONES

What marvels have been reviewed by our dazzled eyes since the outset of these discussions! We first surveyed the magnificent host of stars that people the vast firmament of Heaven; next we admired and wondered at suns very differently constituted from our own; then returning from the depths of space, crossing at a bound the abyss that separates us from these mysterious luminaries, the distant torches of our somber night, terrible suns of infinity, we landed on our own beloved orb, the superb and brilliant day-star. Thence we visited his celestial family, his system, in which our Earth is a floating island. But the journey would be incomplete if we omitted certain more or less vagabond orbs, that occasionally approach the Sun and Earth, some of which may even collide with us upon their celestial path. These are in the first place the comets, then the shooting stars, the fire-balls, and meteorites.

Glittering, swift-footed heralds of Immensity, these comets with golden wings glide lightly through Space, shedding a momentary illumination by their presence. Whence come they? Whither are they bound?

What problems they propound to us, when, as in some beautiful display of pyrotechnics, the arch of Heaven is illuminated with their fantastic light!

But first of all--what is a Comet?

If instead of living in these days of the telescope, of spectrum analysis, and of astral photography, we were anterior to Galileo, and to the liberation of the human spirit by Astronomy, we should reply that the comet is an object of terror, a dangerous menace that appears to mortals in the purity of the immaculate Heavens, to announce the most fatal misfortunes to the inhabitants of our planet. Is a comet visible in the Heavens? The reigning prince may make his testament and prepare to die. Another apparition in the firmament bodes war, famine, the advent of grievous pestilence. The astrologers had an open field, and their fertile imagination might hazard every possible conjecture, seeing that misfortunes, great or small, are not altogether rare in this sublunar world.

How many intellects, and those not the most vulgar, from antiquity to the middle of the last century cursed the apparition of these hirsute stars, which brought desolation to the heart of man, and poured their fatal effluvia upon the head of poor Humanity. The history of the superstitions and fears that they inspired of old would furnish matter for the most thrilling of romances. But, on the other hand, the volume would be little flattering to the common-sense of our ancestors. Despite the respect we owe our forefathers, let us recall for a moment the prejudices attaching to the most famous comets whose passage, as observed from the Earth, has been preserved to us in history.

* * * * *

Without going back to the Deluge, we note that the Romans established a relation between the Great Comet of 43 B.C. and the death of C 鉒 ar, who had been assassinated a few months previously. It was, they asserted, the soul of their great Captain, transported to Heaven to reign in the empyrean after ruling here below. Were not the Emperors Lords of both Earth and Heaven?

We must in justice recognize that certain more independent spirits emancipated themselves from these superstitions, and we may cite the reply of Vespasian to his friends, who were alarmed at the evil presage of a flaming comet: "Fear nothing," he said, "this bearded star concerns me not; rather should it threaten my neighbor the King of the Parthians, since he is hairy and I am bald."

In the year 837 one of these mysterious visitants appeared in the Heavens. It was in the reign of Lewis the Debonair. Directly the King perceived the comet, he sent for an astrologer, and asked what he was to conclude from the apparition. As the answers were unsatisfactory he tried to avert the augury by prayers to Heaven, by ordaining a general fast to all his Court, and by building churches. Notwithstanding, he died three years later, and the historians profited by this slender coincidence to set up a correlation between the fatal star and the death of the Sovereign. This comet, famous in history, is no other than that of Halley, in one of its appearances.

This comet returned to explore the realms near the Sun in 1066, at the moment when William of Normandy was undertaking the Conquest of England, and was misguided enough to go across and reign in London, instead of staying at home and annexing England, thus by his action founding the everlasting rivalry between France and this island. A beneficial influence was attributed to the comet in the Battle of Hastings.

A few centuries later it again came into sight from the Earth, in 1456, three years after the capture of Constantinople by the Turks. Feeling ran high in

Europe, and this celestial omen was taken for a proof of the anger of the Almighty. The moment was decisive; the Christians had to be rescued from a struggle in which they were being worsted. At this conjuncture, Pope Calixtus resuscitated a prayer that had fallen into disuse, the Angelus; and ordered that the bells of the churches should be rung each day at noon, that the Faithful might join at the same hour in prayer against the Turks and the Comet. This custom has lasted down to our own day.

Again, to the comet of 1500 was attributed the tempest that caused the death of Bartholomew Diaz, a celebrated Portuguese navigator, who discovered the Cape of Good Hope.

In 1528 a bearded star of terrific aspect alarmed the world, and the more serious spirits were influenced by this menacing comet, which burned in the Heavens like "a great and gory sword." In a chapter on Celestial Monsters the celebrated surgeon Ambroise Par?describes this awful phenomenon in terms anything but seductive, or reassuring, showing us the menacing sword surrounded by the heads it had cut off (Fig. 50).

Omens of battle, 1547.

Deer and warriors, July 19, 1550.

Cavalry, and a bloody branch crossing the sun, June 11, 1554.]

Our fathers saw many other prodigies in the skies; their descendants, less credulous, can study the facsimile reproduced in Fig. 51, of the drawings published in the year 1557 by Conrad Lycosthenes in his curious Book of Prodigies.

So, too, it is asserted that Charles V renounced the jurisdiction of his Estates, which were so vast that "the Sun never slept upon them," because he was terrified by the comet of 1556 which burned in the skies with an alarming brilliancy, into passing the rest of his days in prayer and devotion.

It is certain that comets often exhibit very strange characteristics, but the imagination that sees in them such dramatic figures must indeed be lively. In the Middle Ages and the Renaissance these were swords of fire, bloody crosses, flaming daggers, etc., all horrible objects ready to destroy our poor human race!

At the time of the Romans, Pliny made some curious distinctions between them: "The Bearded Ones let loose their hair like a majestic beard; the Javelin darts forth like an arrow; if the tail is shorter and ends in a point, it is called the Sword; this is the palest of all the Comets; it shines like a sword, without rays; the Plate or Disk is named in conformity with its figure; its color is amber, the Barrel is actually shaped like a barrel, as it might be in smoke, with light streaming through it; the Horn imitates the figure of a horn erected in the sky, and the Lamp that of a burning flame; the Equine represents a horse's mane, shaken violently with a circular motion. There are bristled comets; these resemble the skins of beasts with the fur on them, and are surrounded by a nebulosity. Lastly, the tails of certain comets have been seen to menace the sky in the form of a lance."

These hairy orbs that appear in all directions, and whose trajectories are sometimes actually perpendicular to the plane of the ecliptic, appear to obey no regular law. Even in the seventeenth century the perspicacious Kepler had not divined their true character, seeing in them, like most of his contemporaries, emanations from the earth, a sort of vapor, losing itself in space. These erratic orbs could not be assimilated with the other members of our grand solar family where, generally speaking, everything goes on in regular order.

And even in our own times, have we not seen the people terrified at the sight of a flaming comet? Has not the end of the world by the agency of comets been often enough predicted? These predictions are so to speak periodic; they crop up each time that the return of these cosmical formations is announced by the astronomers, and always meet with a certain number of timid souls who are troubled as to our destinies.

* * * * *

To-day we know that these wanderers are subject to the general laws that govern the universe. The great Newton announced that, like the planets, they were obedient to universal attraction; that they must follow an extremely elongated curve, and return periodically to the focus of the ellipse. From the basis of these data Halley calculated the progress of the comet of 1682, and ascertained that its motions presented such similarity with the apparitions of 1531 and 1607, that he believed himself justified in identifying them and in announcing its return about the year 1759. Faithful to the call made upon it, irresistibly attracted by the Orb of Day, the comet, at first pale, then ardent and incandescent, returned at the date assigned to it by calculation, three years after the death of the illustrious astronomer. Shining upon his grave it bore witness to the might of human thought, able to snatch the profoundest secrets from the Heavens!

This fine comet returns every seventy-six years, to be visible from the Earth, and has already been seen twenty-four times by the astonished eyes of man. It appears, however, to be diminishing in magnitude. Its last appearance was in 1835, and we shall see it again in 1910, a little sooner than its average period, the attraction of Jupiter having this time slightly accelerated its course, while in 1759 it retarded it.

The comets thus follow a very elongated orbit, either elliptic, turning round the Sun, or parabolic, dashing out into space. In the first case, they are periodic (Fig. 52), and their return can be calculated. In the second they surprise us unannounced, and return to the abysses of eternity to reappear no more.

Their speed is even greater than that of the planets, it is equivalent to this, multiplied by the square root of 2, that is to say by 1.414. Thus at the distance of the Earth from the Sun this velocity = 29,500 meters (18 miles) per second, multiplied by the above number, that is, 41,700 meters (over 25 miles). At the distance of Mercury it = 47 ?1.414 or 66,400 meters (over 40 miles) per second.

Among the numerous comets observed, we do not as yet know more than some twenty of which the orbit has been determined. Periodicity in these bearded orbs is thus exceptional, if we think of the innumerable multitude of comets that circle through the Heavens. Kepler did not exaggerate when he said "there are as many comets in the skies as there are fishes in the sea." These scouts of the sidereal world constitute a regular army, and if we are only acquainted with the dazzling generals clad in gold, it is because the more modest privates can only be detected in the telescope. Long before the invention of the latter, these wanderers in the firmament roamed through space as in our own day, but they defied the human eye, too weak to detect them. Then they were regarded as rare and terrible objects that no one dared to contemplate. To-day they may be counted by hundreds. They have lost in prestige and in originality; but science is the gainer, since she has thus endowed the solar system with new members. No year passes without the announcement of three or four new arrivals. But the fine apparitions that attract general attention by their splendor are rare enough.

These eccentric visitors do not resemble the planets, for they have no opaque body like the Earth, Venus, Mars, or any of the rest. They are transparent nebulosities, of extreme lightness, without mass nor density. We have just photographed the comet of the moment, July, 1903: the smallest stars are visible through its tail, and even through the nucleus.

They arrive in every direction from the depths of space, as though to reanimate themselves in the burning, luminous, electric solar center.

Attracted by some potent charm toward this dazzling focus, they come inquisitive and ardent, to warm themselves at its furnace. At first pale and feeble, they are born again when the Sun caresses them with his fervid heat. Their motions accelerate, they haste to plunge wholly into the radiant light. At length they burst out luminous and superb, when the day-star penetrates them with his burning splendor, illuminates them with a marvelous radiance, and crowns them with glory. But the Sun is generous. Having showered benefits upon these gorgeous celestial butterflies that flutter round him as round some altar of the gods, he grants them liberty to visit other heavens, to seek fresh universes....

The original parabola is converted into an ellipse, if the imprudent adventurer in returning to the Sun passes near some great planet, such as Jupiter, Saturn, Uranus, or Neptune, and suffers its attraction. It is then imprisoned by our system, and can no longer escape from it. After reenforcement at the solar focus, it must return to the identical point at which it felt the first pangs of a new destiny. Henceforward, it belongs to our celestial family, and circles in a closed curve. Otherwise, it is free to continue its rapid course toward other suns and other systems.

* * * * *

As a rule, the telescope shows three distinct parts in a comet. There is first the more brilliant central point, or nucleus, surrounded by a nebulosity called the hair, or brush, and prolonged in a luminous appendix stretching out into the tail. The head of the comet is the brush and the nucleus combined.

It is usually supposed that the tail of a comet follows it throughout the course of its peregrinations. Nothing of the kind. The appendix may even precede the nucleus; it is always opposite the Sun,--that is to say, it is situated on the prolongation of a straight line, starting from the Sun, and passing through the nucleus (Fig. 53). The tail does not exist, so long as the comet is at a distance from the orb of day; but in approaching the Sun, the nebulosity is heated and dilates, giving birth to those mysterious tails and fantastic streamers whose dimensions vary considerably for each comet. The dilations and transformations undergone by the tail suggest that they may be due to a repulsive force emanating from the Sun, an electric charge transmitted doubtless through the ether. It is as though Phoebus blew upon them with unprecedented force.

Telescopic comets are usually devoid of tail, even when they reach the vicinity of the Sun. They appear as pale nebulosities, rounded or oval, more condensed toward the center, without, however, showing any distinct nucleus. These stars are only visible for a minute fraction of their course, when they reach a point not far from the Sun and the terrestrial orbit.

The finest comets of the last century were those of 1811, 1843, 1858, 1861, 1874, 1880, 1881, and 1882. The Great Comet of 1811, after spreading terror over certain peoples, notably in Russia, became the providence of the vine-growers. As the wine was particularly good and abundant that year, the peasants attributed this happy result to the influence of the celestial visitant.

In 1843 one of these strange messengers from the Infinite appeared in our Heavens. It was so brilliant that it was visible in full daylight on February 28th, alongside of the Sun. This splendid comet was accompanied by a marvelous rectilinear tail measuring 300,000,000 kilometers (186,000,000 miles) in length, and its flight was so rapid that it turned the solar hemisphere at perihelion in two hours, representing a speed of 550 kilometers (342 miles) a second.

But the most curious fact is that this radiant apparition passed so near the Sun that it must have traversed its flames, and yet emerged from them safe and sound.

Noteworthy also was the comet of 1858 (Fig. 49), discovered at Florence by Donati. Its tail extended to a length of 90,000,000 kilometers (55,900,000 miles), and its nucleus had a diameter of at least 900 kilometers (559 miles). It is a curious coincidence that the wine was remarkably excellent and abundant in that year also.

The comet of 1861 almost rivaled the preceding.

Coggia's Comet, in 1874, was also remarkable for its brilliancy, but was very inferior to the last two. Finally, the latest worthy of mention appeared in 1882. This magnificent comet also touched the Sun, traveling at a speed of 480 kilometers (299 miles) per second. It crossed the gaseous atmosphere of the orb of day, and then continued its course through infinity. On the day of, and that following, its perihelion, it could be detected with the unaided eye in full daylight, enthroned in the Heavens beside the dazzling solar luminary. For the rest, it was neither that of 1858 nor of 1861.

Since 1882 we have not been favored with a visit from any fine comet; but we are prepared to give any such a reception worthy of their magnificence: first, because now that we have fathomed them we are no longer awestruck; second, because we would gladly study them more closely.

* * * * *

In short, these hirsute stars, whose fantastic appearance impressed the imagination of our ancestors so vividly, are no longer formidable. Their mass is inconsiderable; they seem to consist mainly of the lightest of gases. Analysis of their incandescence reveals a spectrum closely resembling that of many nebul? the presence of carbon is more particularly obvious. Even the nucleus is not solid, and is often transparent.

It is fair to say that the action of a comet might be deleterious if one of these orbs were to arrive directly upon us. The transformation of motion into heat, and the combination of the cometary gases with the oxygen of our atmosphere might produce a conflagration, or a general poisoning of the atmosphere.

But the collision of a comet with a planet is almost an impossibility. This phenomenon could only occur if the comet crossed the planetary orbit at the exact moment at which the planet was passing. When we think of the immensity of space, of the extraordinary length of way traversed by a world in its annual journey round the Sun, and the speed of its rotation, we see why this coincidence is hardly likely to occur. Thus, among the hundreds of comets catalogued, a few only cut the terrestrial orbit. One of them, that of 1832, traversed the path of our globe in the nights of October 29 and 30 in that year; but the Earth only passed the same point thirty days later, and at the critical period was more than 80,000,000 kilometers (50,000,000 miles) away from the comet.

On June 30, 1861, however, the Earth passed through the extremity of the tail of the Great Comet of that year. No one even noticed it. The effects were doubtless quite immaterial.

In 1872 we were to collide with Biela's Comet, lost since 1852; now, as we shall presently see, we came with flying colors out of that disagreeable situation, because the comet had disintegrated, and was reduced to powder. So we may sleep in peace as regards future danger likely to come to us from comets. There is little fear of the destruction of humanity by these windy bags.

These ethereal beauties whose blond locks float carelessly upon the azure night are not concerned with us; they seem to have no other preoccupation than to race from sun to sun, visiting new Heavens, indifferent to the astonishment they produce in us. They speed restlessly and tirelessly through infinity; they are the Amazons of space.

What suns, what worlds must they have visited since the moment of their birth! If these splendid fugitives could relate the story of their wanderings, how gladly should we listen to the enchanting descriptions of the various abodes they have journeyed to! But alas! these mysterious explorers are dumb; they tell none of their secrets, and we must needs respect their enigmatic silence.

Yet, some of them have left us a modest token of remembrance, an almost impalpable nothing, sufficient, however, to enable us to address our thanks to the considerate messenger.

* * * * *

Can there be any one upon the Earth who has not been struck by the phosphorescent lights that glide through the somber night, leaving a brilliant silver or golden track--the luminous, ephemeral trail of a meteor?

Sometimes, when Night has silently spread the immensity of her wings above the weary Earth, a shining speck is seen to detach itself in the shades of evening from the starry vault, shooting lightly through the constellations to lose itself in the infinitude of space.

These bewitching sparks attract our eyes and chain our senses. Fascinating celestial fireflies, their dainty flames dart in every direction through space, sowing the fine dust of their gilded wings upon the fields of Heaven. They are born to die; their life is only a breath; yet the impression which they make upon the imagination of mortals is of the profoundest.

The young girl dreaming in the delicious tranquillity of the transparent night smiles at this charming sister in the Heavens (Fig. 54). What can not this adorable star announce to the tender and loving heart? Is it the shy messenger of the happiness so long desired? Its unpremeditated appearance fills the soul with a ray of hope and makes it tremble. It is a golden beam that glides into the heart, expanding it in the thrills of a sudden and ephemeral pleasure.... The radiant meteor seems to quit the velvet of the deep blue sky to respond to the appeal of the imploring voice that seeks its succor.

What secrets has it not surprised! And who bears malice against it? It is the friend of the betrothed who invoke its passage to confide their wishes, and associate it with their dreams. Tradition holds that if a wish be formulated during the visible passage of a meteor it will certainly be fulfilled before the year is out. Between ourselves, however, this is but a surviving figment of the ancestral imagination, for this celestial jewel takes no such active part in the doings of Humanity.... Besides, try to express a wish distinctly in a second!

It is a curious fact that while comets have so often spread terror on the Earth, shooting stars should on the contrary have been regarded with benevolent feelings at all times. And what is a shooting star? These dainty excursionists from the celestial shores are not, as is supposed, true stars. They are atoms, nothings, minute fragments deriving in general from the disintegration of comets. They come to us from a vast distance, from millions on millions of miles, and circle in swarms around the Sun, following a very elongated ellipse which closely resembles that of the cometary orbit. Their flight is extremely rapid, reaching sometimes more than 40 kilometers (25 miles) per second, a cometary speed that is, as we have seen, greatly above that of our terrestrial vehicle, which amounts to 29 to 30 kilometers (about 19 miles).

These little corpuscles are not intrinsically luminous; but when the orbit of a swarm of meteors crosses our planet, a violent shock arises, the speed of which may be as great as 72 kilometers (45 miles) in the first second if we meet the star shower directly; the average rate, however, does not exceed 30 to 40 kilometers (19 to 25 miles), for these meteors nearly always cross our path obliquely. The height at which they arrive is usually 110 kilometers (68 miles), and 80 kilometers (50 miles) at the moment of disappearance of the meteor; but shooting stars have been observed at 300 kilometers (186 miles).

The friction caused by this collision high up in the atmosphere transforms the motion into heat. The molecules incandesce, and burn like true stars with a brilliancy that is often magnificent.

But their glory is of short duration. The excessive heat resulting from the shock consumes the poor firefly; its remains evaporate, and drop slowly to the Earth, where they are deposited on the surface of the soil in a sort of ferruginous dust mixed with carbon and nickel. Some one hundred and forty-six milliards of them reach us annually, as seen by the unaided eye, and many more in the telescope; the effect of these showers of meteoric matter is an insensible increase in the mass of our globe, a slight lessening of its rotary motion, and the acceleration of the lunar movements of revolution.

Although the appearance of shooting stars is a common enough phenomenon, visible every night of the year, there are certain times when they arrive in swarms, from different quarters of the sky. The most remarkable dates in this connection are the night of August 10th and the morning of November 14th.

Every one knows the shooting stars of August 10th, because they arrive in the fine warm summer evenings so favorable to general contemplation of the Heavens. The phenomenon lasts till the 12th, and even beyond, but the maximum is on the 10th. When the sky is very clear, and there is no moon, hundreds of shooting stars can be counted on those three nights, sometimes thousands. They all seem to come from the same quarter of the Heavens, which is called the radiant, and is situated for the August swarm in the constellation of Perseus, whence they have received the name of Perseids. Our forefathers also called them the tears of St. Lawrence, because the feast of that saint is on the same date. These shooting stars describe a very elongated ellipse, and their orbit has been identified with that of the Great Comet of 1862.

The shower of incandescent asteroids on November 14th is often much more abundant than the preceding. In 1799, 1833, and 1866, the meteors were so numerous that they were described as showers of rain, especially on the first two dates. For several hours the sky was furrowed with falling stars. An English mariner, Andrew Ellicot, who made the drawing we reproduce (Fig. 55), described the phenomenon as stupendous and alarming (November 12, 1799, 3 A.M.). The same occurred on November 13, 1833. The meteors that scarred the Heavens on that night were reckoned at 240,000. These shooting stars received the name of Leonids, because their radiant is situated in the constellation of the Lion.

This swarm follows the same orbit as the comet of 1866, which travels as far as Uranus, and comes back to the vicinity of the Sun every thirty-three years. Hence we were entitled to expect another splendid apparition in 1899, but the expectations of the astronomers were disappointed. All the preparations for the appropriate reception of these celestial visitors failed to bring about the desired result. The notes made in observatories, or in balloons, admitted of the registration of only a very small number of meteors. The maximum was thirteen. During that night, some 200 shooting stars were counted. There were more in 1900, 1901, and, above all, in 1902. This swarm has become displaced.

The night of November 27th again is visited by a number of shooting stars that are the disaggregated remains of the Comet of Biela. This comet, discovered by Biela in 1827, accomplished its revolution in six and a half years, and down to 1846 it responded punctually to the astronomers who expected its return as fixed by calculation. But on January 13, 1846, the celestial wanderer broke in half: each fragment went its own way, side by side, to return within sight from the Earth in 1852. It was their last appearance. That year the twin comets could still be seen, though pale and insignificant. Soon they vanished into the depths of night, and never appeared again. They were looked for in vain, and were despaired of, when on November 27, 1872, instead of the shattered comet, came a magnificent rain of shooting stars. They fell through the Heavens, numerous as the flakes of a shower of snow.

The same phenomenon recurred on November 27, 1885, and confirmed the hypothesis of the demolition and disaggregation of Biela's Comet into shooting stars.

* * * * *

There is an immense variety in the brilliancy of the shooting stars, from the weak telescopic sparks that vanish like a flash of lightning, to the incandescent bolides or fire-balls that explode in the atmosphere.

Fig. 56 shows an example of these, and it represents a fire-ball observed at the Observatory of Juvisy on the night of August 10, 1899. It arrived from Cassiopeia, and burst in Cepheus.

This phenomenon may occur by day as well as by night. It is often accompanied by one or several explosions, the report of which is sometimes perceptible to a considerable distance, and by a shower of meteorites. The globe of fire bursts, and splits up into luminous fragments, scattered in all directions. The different parts of the fire-ball fall to the surface of the Earth, under the name of aerolites, or rather of uranoliths, since they arrive from the depths of space, and not from our atmosphere.

From the most ancient times we hear of showers of uranoliths to which popular superstitions were attached; and the Greeks even gave the name of Sideros to iron, the first iron used having been sidereal.

No year passes without the announcement of several showers of uranoliths, and the phenomenon sometimes causes great alarm to those who witness it. One of the most remarkable explosions is that which occurred above Madrid, February 10, 1896, a fragment from which, sent me by M. Arcimis, Director of the Meteorological Institute, fell immediately in front of the National Museum (Fig. 57). The phenomenon occurred at 9.30 A.M., in brilliant sunshine. The flash of the explosion was so dazzling that it even illuminated the interior of the houses; an alarming clap of thunder was heard seventy seconds after, and it was believed that an explosion of dynamite had occurred. The fire-ball burst at a height of fourteen miles, and was seen as far as 435 miles from Madrid!

In one of Raphael's finest pictures (The Madonna of Foligno) a fire-ball may be seen beneath a rainbow (Fig. 58), the painter wishing to preserve the remembrance of it, as it fell near Milan, on September 4, 1511. This picture dates from 1512.

The dimensions of these meteorites vary considerably; they are of all sizes, from the impalpable dust that floats in the air, to the enormous blocks exposed in the Museum of Natural History in Paris. Many of them weigh several million pounds. That represented below fell in Mexico during the shower of meteors of November 27, 1885. It weighed about four pounds.

These bolides and uranoliths come to us from the depths of space; but they do not appear to have the same origin as the shooting stars. They may arise from worlds destroyed by explosion or shock, or even from planetary volcanoes. The lightest of them may have been expelled from the volcanoes of the Moon. Some of the most massive, in which iron predominates, may even have issued from the bowels of the Earth, projected into space by some volcanic explosion, at an epoch when our globe was perpetually convulsed by cataclysms of extraordinary violence. They return to us to-day after being removed from the Earth to distances proportional to the initial speed imparted to them. This origin seems the more admissible as the stones that fall from the skies exhibit a mineral composition identical with that of the terrestrial materials.

In any case, these uranoliths bring us back at least by their fall to our Earth, and from henceforward we will remain upon it, to study its position in space, and to take account of the place it fills in the Universe, and of the astronomical laws that govern our destiny.

CHAPTER VIII

THE EARTH

Our grand celestial journey lands us upon our own little planet, on this globe that gravitates between Mars and Venus (between War and Love), circulating like her brothers of the solar system, around the colossal Sun.

The Earth! The name evokes in us the image of Life, and calls up the theater of our activities, our ambitions, our joys and sorrows. Does it not, in fact, to ignorant eyes, represent the whole of the universe?

And yet, what is the Earth?

The Earth is a star in the Heavens. We learned this much in our first lesson. It is a globe of opaque material, similar to the planets Mercury, Venus, Mars, Jupiter, etc., as previously described. Isolated on all sides in space, it revolves round the Sun, along a vast orbit that it accomplishes in a year. And while it thus glides along the lines of solar attraction, the terrestrial ball rotates rapidly upon itself in twenty-four hours.

These statements may appear dubious at first sight, and contradictory to the evidence of our senses.

Now that the surface of the Earth has been explored in all directions, there is no longer room to doubt that it is a globe, a sort of ball that we adhere to. A journey round the world is common enough to-day, and always yields the most complete evidence of the spherical nature of the Earth. On the other hand, the curvature of the seas is a no less certain proof. When a ship reaches the dark-blue line that appears to separate the sky from the ocean, it seems to be hanging on the horizon. Little by little, however, as it recedes, it drops below the horizon line; the tops of the masts being the last to disappear. The observer on board ship witnesses the same phenomenon. The low shores are first to disappear, while the high coasts and mountains are much longer visible.

The aspect of the Heavens gives another proof of the Earth's rotundity. As one travels North or South, new stars rise higher and higher above the horizon in the one direction or the other, and those which shine in the latitude one is leaving, gradually disappear. If the surface of the Earth were flat, the ships on the sea would be visible as long as our sight could pierce the distance, and all the stars of the Heavens would be equally visible from the different quarters of the world.

Lastly, during the eclipses of the Moon, the shadow projected by the Earth upon our satellite is always round. This is another proof of the spherical nature of the terrestrial globe.

We described the Earth as an orb in the Heavens, similar to all the other planets of the great solar family. We see these sister planets of our world circulating under the starry vault, like luminous points whose brilliancy is sometimes dazzling. For us they are marvelous celestial birds hovering in the ether, upheld by invisible wings. The Earth is just the same. It is supported by nothing. Like the soap-bubble that assumes a lovely iridescence in the rays of the Sun, or, better, like the balloon rapidly cleaving the air, it is isolated from every kind of support.

Some minds have difficulty in conceiving this isolation, because they form a false notion of weight.

The astronomers of antiquity, who divined it, knew not how to prevent the Earth from falling. They asked anxiously what the strong bands capable of holding up this block of no inconsiderable weight could be. At first they thought it floated on the waters like an island. Then they postulated solid pillars, or even supposed it might turn on pivots placed at the poles. But on what would all these imaginary supports have rested? All these fanciful foundations of the Earth had to be given up, and it was recognized as a globe, isolated in every part. This illusion of the ancients, which still obtains for a great many citizens of our globule, arises, as we said, from a false conception of weight.

Weight and attraction are one and the same force.

A body can only fall when it is attracted, drawn by a more important body. Now, in whatever direction we may wander upon the globe, our feet are always downward. Down is therefore the center of the Earth.

The terrestrial globe may be regarded as an immense ball of magnet, and its attraction holds us at its surface. We weigh toward the center. We may travel over this surface in all directions; our feet will always be below, whatever the direction of our steps. For us, "below" is the inside of our planet, and "above" is the immensity of the Heavens that extend above our heads, right round the globe.

This once understood, where could the Earth fall to? The question is an absurdity. "Below" being toward the center, it would have to fall out of itself.

Let us then picture the Earth as a vast sphere, detached from all that exists around it, in the infinity of the Heavens. A point diametrically opposed to another is called its antipodes. New Zealand is approximately the antipodes to France. Well, for the inhabitants of New Zealand and of France the top is reciprocally opposed, and the bottom, or the feet, are diametrically in opposition. And yet, for one as for the other, the bottom is the soil they are held to, and the top is space above their heads.

The Earth turns on itself in twenty-four hours. Whatever is above us, e.g., at midday, we call high; twelve hours later, at midnight, we give the same qualification to the part of space that was under our feet at noon. What is in the sky, and over our heads, at a given hour, is under our feet, and yet always in the sky, twelve hours later. Our position, in relation to the space that surrounds us, changes from hour to hour, and "top" and "bottom" vary also, relatively to our position.

Our planet is thus a ball, slightly flattened at the poles (by about 1/292). Its diameter, at the equator, is 12,742 kilometers (7,926 miles); from one pole to the other is a little less, owing to the flattening of the polar caps. The difference is some 43 kilometers (about 27 miles).

Its circumference is 40,000 kilometers (24,900 miles). This ball is surrounded by an aerial envelope, the atmosphere, the height of which can not be less than 300 kilometers (186 miles), according to the observations made on certain shooting stars.

We all know that this layer of air, at the bottom of which we live, is a beautiful azure blue that seems to separate us from the sidereal abyss, spreading over our heads in a kind of vault that is often filled with clouds, and giving the illusion of resting far off on the circle of the horizon. But this is only an illusion. In reality, there is neither vault nor horizon; space is open in all directions. If the atmosphere did not exist, or if it were completely transparent, we should see the stars by day as by night, for they are continually round us, at noon as at midnight, and we can see them in the full daylight, with the help of astronomical instruments. In fact, certain stars (the radiant Venus and the dazzling Jupiter) pierce the veil of the atmosphere, and are visible with the unaided eye in full daylight.

The terrestrial surface is 510,000,000 square kilometers (200,000,000 square miles). The waters of the ocean cover three-quarters of this surface, i.e., 383,200,000 square kilometers (150,000,000 square miles), and the continents only occupy 136,600,000 square kilometers (55,000 square miles). France represents about the thousandth part of the total superficies of the globe.

Despite the asperities of mountain ranges, and the abysses hollowed out by the waters, the terrestrial globe is fairly regular, and in relation to its volume its surface is smoother than that of an orange. The highest summits of the Himalaya, the profoundest depths of the somber ocean, do not attain to the millionth part of its diameter.

In weight, the Earth is five and a half times heavier than would be a globe of water of the same dimensions. That is to say:

6,957,930,000,000,000,000,000,000   kilograms
(6,833,000,000,000,000,000,000 tons).

The atmospheric atmosphere with which it is surrounded represents.

6,263,000,000,000,000,000 kilograms (6,151,000,000,000,000 tons).

Each of us carries an average weight of some 17,000 kilograms (16 tons) upon his shoulders. Perhaps some one will ask how it is that we are not crushed by this weight, which is out of all proportion with our strength, but to which, nevertheless, we appear insensible. It is because the aerial fluid enclosed within our bodies exerts a pressure equal and opposite to the external atmospheric pressure, and these pressures are at equilibrium.

The Earth is characterized by no essential or particular differences relatively to the other worlds of our system. Like Venus of the limpid rays, like the dazzling Jupiter, like all the planets, she courses through space, carrying into Infinitude our hopes and destinies. Bigger than Mercury, Venus, and Mars, she presents a very modest figure in comparison with the enormous Jupiter, the strange system of Saturn, of Uranus, and even of Neptune. For us her greatest interest is that she serves as our residence, and if she were not our habitation we should scarcely notice her. Dark in herself, she burns at a distance like a star, returning to space the light she receives from the Sun. At the distance of our satellite, she shines like an enormous moon, fourteen times larger and more luminous than our gentle Phoebe. Observed from Mercury or Venus, she embellishes the midnight sky with her sparkling purity as Jupiter does for us. Seen from Mars, she is a brilliant morning and evening star, presenting phases similar to those which Mars and Venus show from here. From Jupiter, the terrestrial globe is little more than an insignificant point, nearly always swallowed up in the solar rays. As to the Saturnians, Uranians, and Neptunians, if such people exist, they probably ignore our existence altogether. And in all likelihood it is the same for the rest of the universe.

We must cherish no illusions as to the importance of our natal world. It is true that the Earth is not wanting in charm, with its verdant plains enameled in the delicious tones of a robust and varied vegetation, its plants and flowers, its spring-time and its birds, its limpid rivers winding through the meadows, its mountains covered with forests, its vast and profound seas animated with an infinite variety of living creatures. The spectacle of Nature is magnificent, superb, admirable and marvelous, and we imagine that this Earth fills the universe, and suffices for it. The Sun, the Moon, the stars, the boundless Heavens, seem to have been created for us, to charm our eyes and thoughts, to illumine our days, and shed a gentle radiance upon our nights. This is an agreeable illusion of our senses. If our Humanity were extinguished, the other worlds of the Heavens, Venus, Mars, etc., would none the less continue to gravitate in the Heavens along with our defunct planet, and the close of human life (for which everything seems to us to have been created) would not even be perceived by those other worlds, that nevertheless are our neighbors. There would be no revolution, no cataclysm. The stars would go on shining in the firmament, just as they do to-day, shedding their divine light over the immensity of the Heavens. Nothing would be changed in the general aspect of the Universe. The Earth is only a modest atom, lost in the innumerable army of the worlds and suns that people the universe.

* * * * *

Every morning the Sun rises in the East, setting fire with his ardent rays to the sky, which is dazzling with his splendor. He ascends through space, reaches a culminating point at noon, and then descends toward the West, to sink at night into the purple of the sunset.

And then the stars, grand lighthouses of the Heavens, in their turn incandesce. They too rise in the East, ascend the vault of Heaven, and then descend to the West, and vanish. All the orbs, Sun, Moon, planets, stars, appear to revolve round us in twenty-four hours.

This journey of the orbs around us is only an illusion of the senses.

Whether the Earth be at rest, and the sky animated with a rotary movement round her, or whether, on the contrary, the stars are fixed, and the Earth in motion, in either case, for us appearances are the same. If the Earth turns, carrying all that pertains to it in its motion--the seas, the atmosphere, the clouds, and ourselves,--we are unable to perceive it, because all the objects that surround us keep their respective positions among themselves. Hence we must resort to logic, and reason out the two hypotheses.

For the accomplishment of this rapid journey of the Sun and stars around the Earth, it would be necessary that all the orbs of the sky should be in some way attached to a vault, or to circles, as was formerly supposed. This conception is childish. The peoples of antiquity had no notion of the size of the universe, and their error is almost excusable. The distance separating Heaven from the Infernal Regions has been measured, according to Hesiod, by Vulcan's anvil, which fell from the skies to the Earth in nine days and nine nights, and it would have taken as long again to continue its journey from the surface of the Earth to the bowels of Hades.

To-day we have a more exact notion of the grandeur of the Universe. We know that millions and trillions of miles separate the stars from one another. And by representing these distances, we can form some idea of the difficulty there would be in admitting the rotation of the universe round the Earth.

The distance from here to the Sun is 149,000,000 kilometers (93,000,000 miles). In order to turn in twenty-four hours round the Earth, that orb would have to fly through Space at a velocity of more than 10,000 kilometers (6,200 miles) a second.

Yes! the Sun, splendid orb, source of our existence and of that of all the planets, a colossal globe, over a million times more voluminous than the Earth, and 324 thousand times heavier, would have to accomplish this immense revolution in order to turn round the minute point that is our lilliputian world!

This in itself would suffice to convince us of the want of logic in such an argument. But the Sun is not alone in the Heavens. We should have to suppose that all the planets and all the stars were engaged in the same fantastic motions.

Jupiter is about five times as far off as the Sun; his velocity would have to be 53,000 kilometers (32,860 miles) per second.

Neptune, thirty times farther off, would have to execute 320,000 kilometers (198,000 miles) per second.

The nearest star, [alpha] of the Centaur, situated at a distance 275,000 times that of the Sun, would have to run, to fly through space, at a rate of 2,941,000,000 kilometers (1,823,420,000 miles) per second.

All the other stars are incomparably farther off, at infinity.

And this fantastic rotation would all be accomplished round a minute point!

To put the problem in this way is to solve it. Unless we deny the astronomic measures, and the most convincing geometric operations, the Earth's diurnal motion of rotation is a certainty.

To suppose that the stars revolve round the Earth is to suppose, as one author humorously suggests, that in order to roast a pheasant the chimney, the kitchen, the house, and all the countryside must needs turn round it.

If the Earth turns in twenty-four hours upon itself, a point upon the equator would simply travel at a rate of 465 meters (1,525 feet) per second. This speed, while considerable in comparison with the movements observed upon the surface of our planet, is as nothing compared with the fantastic rapidity at which the Sun and stars would have to move, in order to rotate round our globe.

Thus we have to choose between these two hypotheses: either to make the entire Heavens turn round us in twenty-four hours, or to suppose our globe to be animated by a motion of rotation upon itself. For us, the impression is the same, and as we are insensible to the motion of the Earth, its immobility would seem almost natural to us. So that, in last resort, here as in many other instances, the decision must be made by simple common sense. Science long ago made its choice. Moreover, all the progress of Astronomy has confirmed the rotary movement of the Earth in twenty-four hours, and its movement of revolution round the Sun in a year; while at the same time a great number of other motions have been discovered for our wandering planet.

The learned philosophers of antiquity divined the double movement of our planet. The disciples of Pythagoras taught it more than two thousand years ago, and the ancient authors quote among others Nicetas of Syracuse, and Aristarchus of Samos, as being among the first to promote the doctrine of the Earth's movement. But at that remote period no one had any idea of the real distances of the stars, and the argument did not seem to be based on any adequate evidence. Ptolemy, after a long discussion of the diurnal motion of our planet, refutes it, giving as his principal reason that if the Earth turned, the objects that were not fixed to its surface would appear to move in a contrary direction, and that a body shot into the air would fall back to the West of its starting-point, the Earth having turned meantime from West to East. This objection has no weight, because the Earth controls not only all the objects fixed to the soil, but also the atmosphere, and the clouds that surround it like a light veil, and all that exists upon its surface. The atmosphere, the clouds, the waters of the ocean, things and beings, all are adherent to it and make one body with it, participating in its movement, as sometimes happens to ourselves in the compartment of a train, or the car of an aerostat. When, for instance, we drop an object out of such a car, this object, animated with the acquired velocity, does not fall to a point below the aerostat, but follows the balloon, as though it were gliding along a thread. The author has made this experiment more than once in aerial journeys.

Thus, the hypothesis of the Earth's motion has become a certainty. But in addition to reasoning, direct proof is not wanting.

1. The spheroidal shape of the Earth, slightly flattened at the poles and swollen at the equator, has been produced by the rotary motion, by the centrifugal force that it engenders.

2. In virtue of this centrifugal force, which is at its maximum at the equator, objects lose a little of their weight in proportion as they are farther removed from the polar regions where centrifugal force is almost nil.

3. In virtue of this same centrifugal force, the length of the pendulum in seconds is shorter at the equator than in Paris, and the difference is one of 3 millimeters.

4. A weight abandoned to itself and falling from a certain height, should follow the vertical if the Earth were motionless. Experiment, frequently repeated, shows a slight deviation to the East, of the plumb-line that marks the vertical. We more especially observed this at the Pantheon during the recent experiments.

5. The magnificent experiment of Foucault at the Pantheon, just renewed under the auspices of the Astronomical Society of France, demonstrates the rotary motion of the Earth to all beholders. A sufficiently heavy ball (28 kilograms, about 60 pounds) is suspended from the dome of the edifice by an excessively fine steel thread. When the pendulum is in motion, a point attached to the bottom of the ball marks its passage upon two little heaps of sand arranged some yards away from the center. At each oscillation this point cuts the sand, and the furrow gets gradually longer to the right hand of an observer placed at the center of the pendulum. The plane of the oscillations remains fixed, but the Earth revolves beneath, from West to East. The fundamental principle of this experiment is that the plane in which any pendulum is made to oscillate remains invariable even when the point of suspension is turned. This demonstration enables us in some measure to see the Earth turning under our feet.

The annual displacements of the stars are again confirmatory of the Earth's motion round the Sun. During the course of the year, the stars that are least remote from our solar province appear to describe minute ellipses, in perspective, in the Heavens. These small apparent variations in the position of the nearest stars reproduce the annual rotation of the Earth round the Sun, in perspective.

We could adduce further observations in favor of this double movement, but the proofs just given are sufficiently convincing to leave no doubt in the mind of the reader.

Nor are these two the only motions by which our globe is rocked in space. To its diurnal rotation and its annual rotation we may add another series of ten more motions: some very slow, fulfilling themselves in thousands of years, others, more rapid, being constantly renewed. It is, however, impossible in these restricted pages to enter into the detail reserved for more complete works. We must not forget that our present aim is to sum up the essentials of astronomical knowledge as simply as possible, and to offer our readers only the "best of the picking."

* * * * *

The two principal motions of which we have just spoken give us the measure of time, the day of twenty-four hours, and the year of 365-1/4 days.

The Earth turning upon itself in twenty-four hours from West to East, presents all its parts in succession to the Sun fixed in space. Illuminated countries have the day, those opposite, in the shadow of the Earth, are plunged into night. The countries carried by the Earth toward the Sun have morning, those borne toward his shadow, evening. Those which receive the rays of the day-star directly have noon; those which are just opposite have midnight.

The rotation of our planet in this way gives us the measure of time; it has been divided arbitrarily into twenty-four periods called hours; each hour into sixty minutes; each minute into sixty seconds.

In consequence, each country turns in twenty-four hours round the axis of the Earth. The difference in hours between the different regions of the globe is therefore regulated by the difference of geographical position. The countries situated to the West are behind us; the Sun only gets there after it has shone upon our meridian. When it is midday in Paris, it is only 11.51 A.M. in

London; 11.36 A.M. in Madrid; 11.14 A.M. at Lisbon; 11.12 A.M. at Mogador; 7.06 A.M. at Quebec; 6.55 A.M. at New York; 5.14 A.M. in Mexico; and so on. The countries situated to the East are, on the contrary, ahead of us. When it is noon in Paris, it is already 56 minutes after midday at Vienna; 1.25 P.M. at Athens; 2.21 P.M. at Moscow; 3.16 P.M. at Teheran; 4.42 P.M. at Bombay; and so on. We are here speaking of real times, and not of the conventional times.

If we could make the tour of the world in twenty-four hours, starting at midday from some place to go round the globe, and traveling westward with the Sun, we should have him always over our heads. In traveling round the world from West to East, one goes in front of the Sun, and gains by one day; in taking the opposite direction, from East to West, one loses a day.

In reality, the exact duration of the Earth's diurnal rotation is twenty-three hours, fifty-six minutes, four seconds. That is the sidereal day. But, while turning upon itself, the Earth circulates upon its orbit, and at the end of a diurnal rotation it is still obliged to turn during three minutes, fifty-six seconds in order to present exactly the same meridian to the fixed Sun which, in consequence of the rotary period of our planet, is a little behind. The solar day is thus one of twenty-four hours. There are 366 rotations in the year.

And now let us come back to the consequences of the Earth's motion. In the first place our planet does not turn vertically nor on its side, but is tipped or inclined a certain quantity: 23?27'.

Now, throughout its annual journey round the Sun, the inclination remains the same. That is what produces the seasons and climates. The countries which have a larger circle to travel over in the hemisphere of the solar illumination have the longer days, those which have a smaller circle, shorter days. At the equator there is constantly, and all through the year, a twelve-hour day, and a night of twelve hours.

In summer, the pole dips toward the Sun, and the rays of the orb of day cover the corresponding hemisphere with their light. Six months later this same hemisphere is in winter, and the opposite hemisphere is in its turn presented to the Sun. June 21 is the summer solstice for the northern hemisphere, and is at the same time winter for the southern pole. Six months later, on December 21, we have winter, while the southern hemisphere is completely exposed to the Sun. Between these two epochs, when the radiant orb shines exactly upon the equator, that is on March 21, we have the spring equinox, that delicious flowering season when all nature is enchanting and enchanted; on September 21 we have the autumn equinox, melancholy, but not devoid of charm.

The terrestrial sphere has been divided into different zones, with which the different climates are in relation:

1. The tropical zone, which extends 23°27' from one part to the other of the equator. This is the hottest region. It is limited by the circle of the tropics.

2. The temperate zones, which extend from 23°27' to 66°23' of latitude, and where the Sun sets every day.

3. The glacial zones, drawn round the poles, at 66°33' latitude, where the Sun remains constantly above or below the horizon for several days, or even several months. These glacial zones are limited by the polar circles.

We must add that the axis of the Earth is a straight line that is supposed to pass through the center of the globe and come out at two diametrically opposite points called the poles. The diurnal rotation of the Earth is effected round this axis.

The name equator is given to a great circle situated between the two poles, at equal distance, which divides the globe into two hemispheres. The equator is divided into 360 parts or degrees, by other circles that go from one pole to the other. These are the longitudes or meridians (see Fig. 62). The distance between the equator and the pole is divided into larger or smaller circles, which have received the name of latitudes, 90 degrees are reckoned on the one side and the other of the equator, in the direction of the North and South poles, respectively. The longitudes are reckoned from some point either to East or West: the latitudes are reckoned North and South, from the equator. In going from East to West, or inversely, the longitude changes, but in passing from North to South of any spot, it is the latitude that alters.

The circles of latitude are smaller in proportion as one approaches the poles. The circumference of the world is 40,076,600 meters at the equator. At the latitude of Paris (48?50') it is only 26,431,900 meters. A point situated at the equator has more ground to travel over in order to accomplish its rotation in twenty-four hours than a point nearer the pole.

We have already stated that this velocity of rotation is 465 meters per second at the equator. At the latitude of Paris it is not more than 305 meters. At the poles it is nil.

The longitudes, or meridians, are great circles of equal length, dividing the Earth into quarters, like the parts of an orange or a melon. These circumvent the globe, and measure some 40,000,000 (40,008,032) meters. We may remember in passing that the length of the meter has been determined as, by definition, the ten-millionth part of the quarter of a celestial meridian.

Thus, while rotating upon itself, the Earth spins round the Sun, along a vast orbit traced at 149,000,000 kilometers (93,000,000 miles) from the central focus, a sensibly elliptical orbit, as we have already pointed out. It is a little nearer the Sun on January 1st than on July 1st, at its perihelion (peri, near, helios, Sun), than at its aphelion (apo, far, helios, Sun). The difference = 6,000,000 kilometers (3,720,000 miles), and its velocity is a little greater at perihelion than at aphelion.

This second motion produces the year. It is accomplished in three hundred and sixty-five days, six hours, nine minutes, nine seconds. Such is the complete revolution of our planet round the orb of day. It has received the name of sidereal year. But this is not how we calculate the year in practical life. The civil year, known also as the tropical year, is not equivalent to the Earth's revolution, because a very slow gyratory motion, called "the precession of the equinoxes," the cycle of which occupies 25,765 years, drags the spring equinox back some twenty minutes in each year.

The civil year is, accordingly, three hundred and sixty-five days, five hours, forty-eight minutes, forty-six seconds.

In order to simplify the calendar, this accumulating fraction of five hours, forty-eight minutes, forty-six seconds (about a quarter day) is added every four years to a bissextile year (leap-year), and thus we have uneven years of three hundred and sixty-five, and three hundred and sixty-six days. Every year of which the figure is divisible by four is a leap-year. By adding a quarter day to each year, there is a surplus of eleven minutes, fourteen seconds. These are subtracted every hundred years by not taking as bissextile those secular years of which the radical is not divisible by four. The year 1600 was leap-year: 1700, 1800, and 1900 were not; 2000 will be. The agreement between the calendar and nature has thus been fairly perfect, since the establishment of the Gregorian Calendar in 1582.

Since the terrestrial orbit measures not less than 930,000,000 kilometers (576,600,000 miles), which must be traversed in a year, the Earth flies through Space at 2,544,000 kilometers (1,577,280 miles) a day, or 106,000 kilometers (65,720 miles) an hour, or 29,500 meters (18 miles) per second on an average, a little faster at perihelion, a little slower at aphelion. This giddy course, a thousand times more rapid than the speed of an express-train, is effected without commotion, shock, or noise. Reasoning alone enables us to divine the prodigious movement that carries us along in the vast fields of the Infinite, in mid-heaven.

Returning to the calendar, it must be remarked in conclusion, that the human race has not exhibited great sense in fixing the New Year on January 1. No more disagreeable season could have been selected. And further, as the ancient Roman names of the months have been preserved, which in the time of Romulus began with March, the "seventh" month, "September," is our ninth month; October (the eighth) is the tenth; November (the ninth) has become the eleventh; and December (the tenth) has taken the place of the twelfth. Verily, we are not hard to please!

These months, again, are unequal, as every one knows. Witness the simple expedient of remembering the long and short months, by closing the left hand and counting the knobs and hollows of the fist, the former corresponding to the long months, the latter to the short: first knob = January; first hollow, February; second knob, March; and so on.[12]

Should not the real renewal of the year coincide with the awakening of Nature, with the spring on the terrestrial hemisphere occupied by the greater portion of Humanity, with the date of March 21st? Should not the months be equalized, and their names modified? Why should we not follow the beautiful evolution dictated by the Sun and by the movement of our planet? But our poor Earth may roll on a long time yet before its inhabitants will become reasonable.

CHAPTER IX

THE MOON

It is the delightful hour when all Nature pauses in the tranquil calm of the silent night.

The Sun has cast his farewell gleams upon the weary Earth. All sound is hushed. And soon the stars will shine out one by one in the bosom of the somber firmament. Opposite to the sunset, in the east, the Full Moon rises slowly, as it were calling our thoughts toward the mysteries of eternity, while her limpid night spreads over space like a dew from Heaven.

In the odorous woods, the trees are silhouetted strangely upon the sky, seeming to stretch their knotted arms toward this celestial beauty. On the river, smooth as a mirror, wherein the pale Phoebe reflects her splendor, the maidens go to seek the floating image of their future spouse. And in response to their prayers, she rends the veil of cloud that hides her from their eyes, and pours the reflection of her gentle beams upon the sleeping waters.

From all time the Moon has had the privilege of charming the gaze, and attracting the particular attention of mortals. What thoughts have not been wafted to her pale, yet luminous disk? Orb of mystery and of solitude, brooding over our silent nights, this celestial luminary is at once sad and splendid in her glacial purity, and her limpid rays provoke a reverie full of charm and melancholy. Mute witness of terrestrial destinies, her nocturnal flame watches over our planet, following it in its course as a faithful satellite.

The human eye first uplifted to the Heavens was struck, above all, with the brilliancy of this solitary globe, straying among the stars. The Moon first suggested an easy division of time into months and weeks, and the first astronomical observations were limited to the study of her phases.

Daughter of the Earth, the Moon was born at the limits of the terrestrial nebula, when our world was still no more than a vast gaseous sphere, and was detached from her at some critical period of colossal solar tide. Separating with regret from her cradle, but attached to the Earth by indissoluble ties of attraction, she rotates round us in a month, from west to east, and this movement keeps her back a little each day in relation to the stars. If we watch, evening by evening, beginning from the new moon, we shall observe that she is each night a little farther to the left, or east, than on the preceding evening. This revolution of the Moon around our planet produces the phases, and gives the measure of our months.

During her monthly journey she always presents the same face to us. One might think that the fear of losing us had immobilized her globe, and prevented her from turning. And so we only know of her the vague sketch of a human face that has been observed through all the ages.

It seems, in fact, as though she were looking down upon us from the Heavens, the more so as the principal spots of her disk vaguely recall the aspect of a face. If we try to draw it without the aid of instruments we observe dark regions and clear regions that each interprets in his own fashion. To the author, for instance, the full Moon has the appearance represented in the following figure. The spots resemble two eyes and the sketch of a nose; resulting in a vague human figure, as indicated on the lower disk. Others see a man carrying a bundle of wood, a hare, a lion, a dog, a kangaroo, a sickle, two heads embracing, etc.[13] But generally speaking, there is a tendency to see a human figure in it.

If this appearance is helped a little by drawing, it gives the profile of a man's head fairly well sketched, and furnished with an abundant crop of hair (Fig. 66). Others go much more into detail, and draw a woman's head that is certainly too definite, like this of M. Jean Sardou (Fig. 67). Others, again, like M. Zamboni, see behind the man's profile the likeness of a young girl being embraced by him (Fig. 68). There is certainly some imagination about these. And yet, on the first suitable occasion, look at the Moon through an opera-glass, a few days after the first quarter, and you will not fail to see the masculine profile just described, and even to imagine the "kiss in the Moon."

These vague aspects disappear as soon as the Moon is examined with even the least powerful instruments: the spots are better defined, and the illusions of indistinct vision vanish. Compare this direct photograph of the Moon, taken by the author some years ago (Fig. 69): here is neither a human figure, man, dog, hare, nor faggot; simply deep geographical configurations, and in the lower region, a luminous point whence certain light bands spread out, some being prolonged to a considerable distance. And yet, from a little way off, does it not form the man's face above indicated?

From the earliest astronomical observations made with the aid of instruments by Galileo, in 1609, people tried to find out what the dark spots could represent, and they were called seas, because water absorbs light, and reflects it less than terra firma. The Moon of itself possesses no intrinsic light, any more than our planet, and only shines by the light of the Sun that illuminates it. As it rotates round the Earth, and constantly changes its position with respect to the Sun, we see more or less of its illuminated hemisphere, and the result is the phases that every one knows so well.

At the commencement of each lunation, the Moon is between the Sun and the Earth, and its non-illuminated hemisphere is turned toward us. This is the New Moon, invisible to us; but two days later, the slim crescent of Diana sheds a gentle radiance upon the Earth. Gradually the crescent enlarges. When the Moon arrives at right angles with ourselves and with the Sun, half the illuminated hemisphere is presented to us. This is the first quarter. At the time of Full Moon, it is opposite the Sun, and we see the whole of the hemisphere illuminated. Then comes the decline: the brilliant disk is slightly corroded at first; it diminishes from day to day, and about a week before the New Moon our fair friend only shows her profile before she once more passes in front of the Sun: this is the last quarter.

When the Moon is crescent, in the first evenings of the lunation, and after the last quarter, the rest of the disk is visible, illuminated feebly by a pale luminosity. This is known as the ashy light. It is due to the shine of the Earth, reflecting the light received from the Sun into space. Accordingly the ashy light is the reflection of our own sent back to us by the Moon. It is the reflection of a reflection.

This rotation of the Moon round the Earth is accomplished in twenty-seven days, seven hours, forty-three minutes, eleven seconds; but as the Earth is simultaneously revolving round the Sun, when the Moon returns to the same point (the Earth having become displaced relatively to the Sun), the Moon has to travel two days longer to recover its position between the Sun and the Earth, so that the lunar month is longer than the sidereal revolution of the Moon, and takes twenty-nine days, twelve hours, forty-four minutes, three seconds. This is the duration of the sequence of phases.

This revolution is accomplished at a distance of 384,000 kilometers (238,000 miles). The velocity of the Moon in its orbit is more than 1 kilometer (0.6214 mile) per second. But our planet sweeps it through space at a velocity almost thirty times greater.

The diameter of the Moon represents 273/1000 that of the Earth, i.e., 3,480 kilometers (2,157 miles).

Its surface = 38,000,000 square kilometers (15,000,000 square miles), a little more than the thirteenth part of the terrestrial surface, which = 510,000,000 (200,000,000 square miles).

In volume, the Moon is fifty times less than the Earth. Its mass or weight is only 1/81 that of the terrestrial globe. Its density = 0.615, relatively to that of the Earth, i.e., a little more than three times that of water. Weight at its surface is very little: 0.174. A kilogram transported thither would only weigh 174 grams.

* * * * *

At the meager distance of 384,000 kilometers (238,000 miles) that separates us from it (about thirty times the diameter of the Earth), the Moon is a suburb of our terrestrial habitation. What does this small distance amount to? It is a mere step in the universe.

A telegraphic message would get there in one and a half second; a projectile fired from a gun would arrive in eight days, five hours; an express-train would be due in eight months, twenty-two days. It is only the 1/388 part of the distance that separates us from the Sun, and only the 100/1,000,000 part of the distance of the stars nearest to us. Many men have tramped the distance that separates us from the Moon. A bridge of thirty terrestrial globes would suffice to unite the two worlds.

Owing to this great proximity, the Moon is the best known of all the celestial spheres. Its geographical (or more correctly, selenographical, Selene, moon) map was drawn out more than two centuries ago, at first in a vague sketch, and afterward with more details, until to-day it is as precise and accurate as any of our terrestrial maps of geography.

Before the invention of the telescope, from antiquity to the seventeenth century, people lost themselves in conjectures as to the nature of this strange lunar figure. It was held to be a mysterious world, the more extraordinary in that it always presented the same face to us. Some compared it to an immense mirror reflecting the image of the Earth. Others pictured it as a silver star, an enchanted abode where all was wealth and happiness. For many a long day it was the fashion to think, quite irrationally, that the inhabitants of the Moon were fifteen times bigger than ourselves.

The invention of telescopes, however, brought a little order and a grain of truth into these fantastic assumptions. The first observations of Galileo revolutionized science, and his discoveries filled the best-ordered minds with enthusiasm. Thenceforward, the Moon became our property, a terrestrial suburb, where the whole world would gladly have installed itself, had the means of getting there been as swift as the wings of the imagination. It became easy enough to invent a thousand enchanting descriptions of the charms of our fair sister, and no one scrupled to do so. Soon, it was observed that the Moon closely resembled the Earth in its geological features; its surface bristles with sharp mountain peaks that light up in so many luminous points beneath the rays of the Sun. Alongside, dark and shaded parts indicate the plains; moreover, there are large gray patches that were supposed to be seas because they reflect the solar light less perfectly than the adjacent countries. At that epoch hardly anything was known of the physical constitution of the Moon, and it was figured as enveloped with an atmospheric layer, analogous to that at the bottom of which we carry on our respiration.

To-day we know that these "seas" are destitute of water, and that if the lunar globe possesses an atmosphere, it must be excessively light.

The Moon became the favorite object of astronomers, and the numerous observations made of it authorized the delineation of very interesting selenographic charts. In order to find one's way among the seas, plains, and mountains that make up the lunar territory, it was necessary to name them. The seas were the first to be baptized, in accordance with their reputed astrological influences. Accordingly, we find on the Moon, the Sea of Fecundity, the Lake of Death, the Sea of Humors, the Ocean of Tempests, the Sea of Tranquillity, the Marsh of Mists, the Lake of Dreams, the Sea of Putrefaction, the Peninsula of Reverie, the Sea of Rains, etc.

With regard to the luminous parts and the mountains, it was at first proposed to call them after the most illustrious astronomers, but the fear of giving offense acted as a check on Hevelius and Riccioli, authors of the first lunar maps (1647, 1651), and they judged it more prudent to transfer the names of the terrestrial mountains to the Moon. The Alps, the Apennines, the Pyrenees, the Carpathians, are all to be found up there; then, as the vocabulary of the mountains was not adequate, the scientists reasserted their rights, and we meet in the Moon, Aristotle, Plato, Hipparchus, Ptolemy, Copernicus, Kepler, Newton, as well as other more modern and even contemporaneous celebrities.

We have not space to reproduce the general chart of the Moon (that published by the author measures not less than a meter, with the nomenclature); but the figure subjoined gives a summary sufficient for the limits of this little book. Here are the names of the principal lunar mountains, with the numbers corresponding to them upon the map.

1 Furnerius 2 Petavius 3 Langrenus 4 Macrobius 5 Cleomedes 6 Endymion 7 Altas 8 Hercules 9 Romer 10 Posidonius 11 Fracastorius 12 Theophilus 13 Piccolomini 14 Albategnius 15 Hipparchus 16 Manilius 17 Eudoxus 18 Aristotle 19 Cassini 20 Aristillus 21 Plato 22 Archimedes 23 Eratosthenes 24 Copernicus 25 Ptolemy 26 Alphonsus 27 Arzachel 28 Walter 29 Clavius 30 Tycho 31 Bullialdus 32 Schiller 33 Schickard 34 Gassendi 35 Kepler 36 Grimaldi 37 Aristarchus

A Mare Crisum B Mare Fercunditatis C Mare Nectaris D Mare Tranquilitatis E Mare Serenitatis F Mare Imbrium G Sinus Iridum H Oceanus Procellarum I Mare Humorum K Mare Nubium V Altai Mountains W Mare Vaporum X Apennine Mountains Y Caucasus Mountains Z Alps]

The constantly growing progress of optics leads to perpetual new discoveries in science, and at the present time we can say that we know the geography of the Moon as well as, and even better than, that of our own planet. The heights of all the mountains of the Moon are measured to within a few feet. (One cannot say as much for the mountains of the Earth.) The highest are over 7,000 meters (nearly 25,000 feet). Relatively to its proportions, the satellite is much more mountainous than the planet, and the plutonian giants are much more numerous there than here. If we have peaks, like the Gaorisankar, the highest of the Himalayas and of the whole Earth, whose elevation of 8,840 meters (29,000 feet) is equivalent to 1/1140 the diameter of our globe, there are peaks on the Moon of 7,700 meters (25,264 feet), e.g., those of Doerfel and Leibniz, the height of which is equivalent to 1/470 the lunar diameter.

Tycho's Mountain is one of the finest upon our satellite. It is visible with the naked eye (and perfectly with opera-glasses) as a white point shining like a kind of star upon the lower portion of the disk. At the time of full moon it is dazzling, and projects long rays from afar upon the lunar globe. So, too, Mount Copernicus, whose brilliant whiteness sparkles in space. But the strangest thing about these lunar mountains is that they are all hollow, and can be measured as well in depth as in height. A type of mountain as strange to us as are the seas without water! In effect, these mountains of the moon are ancient volcanic craters, with no summits, nor covers.

At the top of the highest peaks, there is a large circular depression, prolonged into the heart of the mountain, sometimes far below the level of the surrounding plains, and as these craters often measure several hundred kilometers, one is obliged, if one does not want to go all round them in crossing the mountain, to descend almost perpendicularly into the depths and cross there, to reascend the opposite side, and return to the plain. These alpine excursions incontestably deserve the name of perilous ascents!

No country on the Earth can give us any notion of the state of the lunar soil: never was ground so tormented; never globe so profoundly shattered to its very bowels. The mountains are accumulations of enormous rocks tumbled one upon the other, and round the awful labyrinth of craters one sees nothing but dismantled ramparts, or columns of pointed rocks like cathedral spires issuing from the chaos.

As we said, there is no atmosphere, or at least so little at the bottom of the valleys that it is imperceptible. No clouds, no fog, no rain nor snow. The sky is an eternally black space, vaultless, jeweled with stars by day as by night.

Let us suppose that we arrive among these savage steppes at daybreak: the lunar day is fifteen times longer than our own, because the Sun takes a month to illuminate the entire circuit of the Moon; there are no less than 354 hours from the rising to the setting of the Sun. If we arrive before the sunrise, there is no aurora to herald it, for in the absence of atmosphere there can be no sort of twilight. Of a sudden on the dark horizon come flashes of the solar light, striking the summits of the mountains, while the plains and valleys are still in darkness. The light spreads slowly, for while on the Earth in central latitudes the Sun takes only two minutes and a quarter to rise, on the Moon it takes nearly an hour, and in consequence the light it sends out is very weak for some minutes, and increases excessively slowly. It is a kind of aurora, but lasts a very short time, for when at the end of half an hour, the solar disk has half risen, the light appears as intense to the eye as when it is entirely above the horizon; the radiant orb is seen with its protuberances and its burning atmosphere. It rises slowly, like a luminous god, in the depths of the black sky, a profound and formless sky in which the stars shine all day, since they are not hidden by any atmospheric veil such as conceals them from us during the daylight.

The absence of sensible atmosphere must produce an effect on the temperature of the Moon analogous to that perceived on the high mountains of our globe, where the rarefaction of the air does not permit the solar heat to concentrate itself upon the surface of the soil, as it does below the atmosphere, which acts as a forcing-house: the Sun's heat is not kept in by anything, and incessantly radiates out toward space. In all probability the cold is extremely and constantly rigorous, not only during the nights, which are fifteen times longer than our own, but even during the long days of sunshine.

We give two different drawings to represent these curious aspects of lunar topography. The first (Fig. 72) is taken in the neighborhood of the Apennines, and shows a long chain of mountains beneath which are three deep rings, Archimedes, Aristillus, and Autolycus: the second (Fig. 73) depicts the lunar ring of Flammarion,[14] whose outline is constructed of dismantled ramparts, and whose depths are sprinkled with little craters. The first of these two drawings was made in England by Nasmyth, the second in Germany by Krieger: they both give an exact idea of what one sees in the telescope with different modes of solar illumination.

In the Moon's always black and starry sky a majestic star that is not visible from the Earth, and exhibits this peculiarity that it is stationary in the Heavens, while all the others pass behind it, may constantly be admired, by day as well as by night; and it is also of considerable apparent magnitude. This orb, some four times as large as the Moon in diameter, and thirteen to fourteen times more extensive in surface, is our Earth, which presents to the Moon a sequence of phases similar to those which our satellite presents to us, but in the inverse direction. At the moment of New Moon, the Sun fully illuminates the terrestrial hemisphere turned toward our satellite, and we get "Full Earth"; at the time of Full Moon, on the contrary, the non-illuminated hemisphere of the Earth is turned toward the satellite, and we get "New Earth": when the Moon shows us first quarter, the Earth is in last quarter, and so on. The drawing subjoined gives an idea of these aspects.

What a curious sight our globe must be during this long night of fourteen times twenty-four hours! Independent of its phases, which bring it from first quarter to full earth for the middle of the night, and from full earth to last quarter for sunrise, how interested we should be to see it thus stationary in the sky, and turning on itself in twenty-four hours.

Yes, thanks to us, the inhabitants of the lunar hemisphere turned toward us are gratified by the sight of a splendid nocturnal torch, doubtless less white than our own despite the clouds with which the terrestrial globe is studded, and shaded in a tender tone of bluish emerald-green. The royal orb of their long nights, the Earth, gives them moonlight of unparalleled beauty, and we may say without false modesty that our presence in the lunar sky must produce marvelous and absolutely fairy-like effects.

Maybe, they envy us our globe, a dazzling dwelling-place whose splendor radiates through space; they see its greenish clarity varying with the extent of cloud that veils its seas and continents, and they observe its motion of rotation, by which all the countries of our planet are revealed in succession to its admirers.

We are talking of these pageants seen from the Moon, and of the inhabitants of our satellite as if they really existed. The sterile and desolate aspect of the lunar world, however, rather brings us to the conclusion that such inhabitants are non-existent, although we have no authorization for affirming this. That they have existed seems to me beyond doubt. The lunar volcanoes had a considerable activity, in an atmosphere that allowed the white volcanic ashes to be carried a long way by the winds, figuring round the craters the stellar rays that are still so striking. These cinders were spread over the soil, preserving all its asperities of outline, a little heaped up on the side to which they were impelled. The magnificent photographs recently made at the Paris Observatory by MM. Loewy and Puiseux are splendid evidence of these projections. In this era of planetary activity there were liquids and gases on the surface of the lunar globe, which appear subsequently to have been entirely absorbed. Now the teaching of our own planet is that Nature nowhere remains infertile, and that the production of Life is a law so general and so imperious that life develops at its own expense, sooner than abstain from developing. Accordingly, it is difficult to suppose that the lunar elements can have remained inactive, when only next door they exhibited such fecundity upon our globe. Yes, the Moon has been inhabited by beings doubtless very different from ourselves, and perhaps may still be, although this globe has run through the phases of its astral life more rapidly than our own, and the daughter is relatively older than the mother.

The duration of the life of the worlds appears to have been in proportion with their masses. The Moon cooled and mineralized more quickly than the Earth. Jupiter is still fluid.

The progress of optics brings us already very close to this neighboring province. 'Tis a pity we can not get a little nearer!

A telescopic magnification of 2,000 puts the Moon at 384,000/2000 or 192 kilometers (some 120 miles) from our eye. Practically we can obtain no more, either from the most powerful instruments, or from photographic enlargements. Sometimes, exceptionally, enlargements of 3,000 can be used. This = 384000/3000 or 128 kilometers (some 80 miles). Undoubtedly, this is an admirable result, which does the greatest honor to human intelligence. But it is still too far to enable us to determine anything in regard to lunar life.

Any one who likes to be impressed by grand and magnificent sights may turn even a modest field-glass upon our luminous satellite, at about first quarter, when the relief of its surface, illuminated obliquely by the Sun, is at its greatest value. If you examine our neighbor world at this period, for choice at the hour of sunset, you will be astonished at its brilliancy and beauty. Its outlines, its laces, and embroideries, give the image of a jewel of shining silver, translucent, fluid, palpitating in the ether. Nothing could be more beautiful, nothing purer, and more celestial, than this lunar globe floating in the silence of space, and sending back to us as in some fairy dream the solar illumination that floods it. But yesterday I received the same impression, watching a great ring half standing out, and following the progress of the Sun as it mounted the lunar horizon to touch these silvered peaks. And I reflected that it is indeed inconceivable that 999,999/1,000,000 of the inhabitants of our planet should pass their lives without ever having attended to this pageant, nor to any of those others which the divine Urania scatters so profusely beneath the wondering gaze of the observers of the Heavens.

CHAPTER X

THE ECLIPSES

Among all the celestial phenomena at which it may be our lot to assist during our contemplation of the universe, one of the most magnificent and imposing is undoubtedly that which we are now going to consider.

The hirsute comets, and shooting stars with their graceful flight, captivate us with a mysterious and sometimes fantastic attraction. We gladly allow our thoughts, mute questioners of the mysteries of the firmament, to rest upon the brilliant, golden trail they leave behind them. These unknown travelers bring a message from eternity; they tell us the tale of their distant journeys. Children of space, their ethereal beauty speaks of the immensity of the universe.

The eclipses, on the other hand, are phenomena that touch us more nearly, and take place in our vicinity.

In treating of them, we remain between the Earth and the Moon, in our little province, and witness the picturesque effects of the combined movements of our satellite around us.

Have you ever seen a total eclipse of the Sun?

The sky is absolutely clear: no fraction of cloud shadows the solar rays. The azure vault of the firmament crowns the Earth with a dome of dazzling light. The fires of the orb of day shed their beneficent influence generally upon the world.

Yet, see! The radiance diminishes. The luminous disk of the Sun is gradually corroded. Another disk, as black as ink, creeps in front of it, and little by little invades it entirely. The atmosphere takes on a wan, sepulchral hue; astonished nature is hushed in profound silence; an immense veil of sadness spreads over the world. Night comes on suddenly, and the stars shine out in the Heavens. It seems as though by some mysterious cataclysm the Sun had disappeared forever. But this tribulation is soon over. The divine orb is not extinct. A flaming jet emerges from the shadow, announcing his return, and when he reappears we see that he has lost nothing in splendor or beauty. He is still the radiant Apollo, King of Day, watching over the life of the planetary worlds.

This sudden night, darkening the Heavens in the midst of a fine day, can not fail to produce a vivid impression upon the spectators of the superb phenomenon.

The eclipse lasts only for a few moments, but long enough to make a deep impression upon our minds, and indeed to inspire anxious spirits with terror and agitation--even at this epoch, when we know that there is nothing supernatural or formidable about it.

In former days, Humanity would have trembled, in uneasy consternation. Was it a judgment from Heaven? Must it not be the work of some invisible hand throwing the somber veil of night over the celestial torch?

Had not the Earth strayed off her appointed path, and were we not all to be deprived eternally of the light of our good Sun? Was some monstrous dragon perhaps preparing to devour the orb of day?

The fable of the dragon devouring the Sun or Moon during the eclipses is universal in Asia as in Africa, and still finds acceptance under more than one latitude. But our readers already know that we may identify the terrible celestial dragon with our gentle friend the Moon, who would not be greatly flattered by the comparison.

We saw in the preceding lesson that the Moon revolves round us, describing an almost circular orbit that she travels over in about a month. In consequence of this motion, the nocturnal orb is sometimes between the Sun and the Earth, sometimes behind us, sometimes at a right angle in relation to the Sun and the Earth. Now, the eclipses of the Sun occur invariably at the time of New Moon, when our satellite passes between the Sun and ourselves, and the eclipses of the Moon, at the moment of Full Moon, when the latter is opposite to the Sun, and behind us.

This fact soon enabled the astronomers of antiquity to discover the causes to which eclipses are due.

The Moon, passing at the beginning of its revolution between the Sun and the Earth, may conceal a greater or lesser portion of the orb of day. In this case there is an eclipse of the Sun. On the other hand, when it is on the other side of the Earth in relation to the Sun, at the moment of Full Moon, our planet may intercept the solar rays, and prevent them from reaching our satellite. The Moon is plunged into the shadow of the Earth, and is then eclipsed. Such is the very simple explanation of the phenomenon. But why is there not an eclipse of the Sun at each New Moon, and an eclipse of the Moon at each Full Moon?

If the Moon revolved round us in the same plane as the Earth round the Sun, it would eclipse the Sun at each New Moon, and would be itself eclipsed in our shadow at each Full Moon. But the plane of the lunar orbit dips a little upon the plane of the terrestrial orbit, and the eclipses can only be produced when the New Moon or the Full Moon occur at the line of intersection of these two planes, i.e., when the Sun, the Moon, and the Earth are upon the same straight line. In the majority of cases, instead of interposing itself directly in front of the sovereign of our system, our satellite passes a little above or a little below him, just as its passage behind us is nearly always effected a little above or below the cone of shadow that accompanies our planet, opposite the Sun.

When the Moon intervenes directly in front of the Sun, she arrests the light of the radiant orb, and conceals a greater or less portion of the solar disk. The eclipse is partial if the Moon covers only a portion of the Sun; total if she covers it entirely; annular, if the solar disk is visible all round the lunar disk, as appears when the Moon, in her elliptical orbit, is beyond medium distance, toward the apogee.

On the other hand, when the Moon arrives immediately within the cone of shadow that the Earth projects behind it, it is her turn to be eclipsed. She no longer receives the rays of the Sun, and this deprivation is the more marked in that she owes all her brilliancy to the light of the orb of day. The Moon's obscurity is complete if she is entirely plunged into the cone of shadow. In this case, the eclipse is total. But if a portion of her disk emerges from the cone, that part remains illuminated while the light of the other dies out. In that case there is a partial eclipse, and the rounded form of the Earth's shadow can be seen projected upon our satellite, a celestial witness to the spherical nature of our globe.

Under certain conditions, then, the Moon can deprive us of the luminous rays of the Sun, by concealing the orb of day, and in other cases is herself effaced in crossing our shadow. Despite the fables, fears, and anxieties it has engendered, this phenomenon is perfectly natural: the Moon is only playing hide-and-seek with us--a very harmless amusement, as regards the safety of our planet.

But as we said just now, these phenomena formerly had the power of terrifying ignorant mortals, either when the orb of light and life seemed on the verge of extinction, or when the beautiful Phoebus was covered with a veil of crape and woe, or took on a deep coppery hue.

It would take a volume to describe all the notable events which have been influenced by eclipses, sometimes for good, more often with disastrous consequences. The recital of these tragic stories would not be devoid of interest; it would illustrate the possibilities of ignorance and superstition, and the power man gains from intellectual culture and scientific study.

Herodotus records that the Scythians, having some grievance against Cyaxarus, King of the Medes, revenged themselves by serving up the limbs of one of his children, whom they had murdered, at a banquet as rare game. The scoundrels who committed this atrocious crime took refuge at the Court of the King of Lydia, who was ill judged enough to protect them. War was accordingly declared between the Medes and Lydians, but a total eclipse of the

Sun occurring just when the battle was imminent, had the happy effect of disarming the combatants, who prudently retired each to their own country. This eclipse, which seems to have occurred on May 28, 584 B.C., had been predicted by Thales. The French painter Rochegrosse has painted a striking picture of the scene (Fig. 75).

In the year 413 B.C. the Athenian General Nicias prepared to return to Greece after an expedition to Sicily. But, terrified by an eclipse of the Moon, and fearing the malign influence of the phenomenon, he put off his departure, and lost the chance of retreat. This superstition cost him his life. The Greek army was destroyed, and this event marks the commencement of the decadence of Athens.

In 331 B.C. an eclipse of the Moon disorganized the troops of Alexander, near Arbela, and the great Macedonian Captain had need of all his address to reassure his panic-stricken soldiers.

Agathocles, King of Syracuse, blocked by the Carthaginians in the port of this city, had the good fortune to escape, but was disturbed on the second day of his flight by the arrival of a total eclipse of the Sun which alarmed his companions. "What are you afraid of?" said he, spreading his cloak in front of the Sun. "Are you alarmed at a shadow?" (This eclipse seems to be that of August 15, 309, rather than that of March 2, 310.)

[Illustration: FIG. 75.--Battle between the Medes and Lydians arrested by an Eclipse of the Sun.]

On June 29, 1033, an epoch at which the approaching end of the world struck terror into all hearts, an annular eclipse of the Sun occurring about midday frustrated the designs of a band of conspirators who intended to strangle the Pope at the altar. This Pope was Benedict IX, a youth of less than twenty, whose conduct is said to have been anything but exemplary. The assassins, terrified at the darkening of the Sun, dared not touch the Pontiff, and he reigned till 1044.[15]

On March 1, 1504, a lunar eclipse saved the life of Christopher Columbus. He was threatened with death by starvation in Jamaica, where the contumacious savages refused to give him provisions. Forewarned of the arrival of this eclipse by the astronomical almanacs, he threatened to deprive the Caribs of the light of the Moon--and kept his word. The eclipse had hardly begun when the terrified Indians flung themselves at his feet, and brought him all that he required.

In all times and among all people we find traces of popular superstitions connected with eclipses. Here, the abnormal absence of the Moon's light is regarded as a sign of divine anger: the humble penitents betake themselves to prayer to ward off the divine anger. There, the cruelty of the dread dragon is to be averted: he must be chased away by cries and threats, and the sky is bombarded with shots to deliver the victim from his monstrous oppressor.

In France the announcement of a solar eclipse for August 21, 1560, so greatly disturbed our ancestors' peace of mind as to make them idiotic. Preparations were made for assisting at an alarming phenomenon that threatened Humanity with deadly consequences! The unhappy eclipse had been preceded by a multitude of ill omens! Some expected a great revolution in the provinces and in Rome, others predicted a new universal deluge, or, on the other hand, the conflagration of the world; the most optimistic thought the air would be contaminated. To preserve themselves from so many dangers, and in accordance with the physicians' orders, numbers of frightened people shut themselves up in tightly closed and perfumed cellars, where they awaited the decrees of Fate. The approach of the phenomenon increased the panic, and it is said that one village cur? being unable to hear the confessions of all his flock, who wanted to discharge their souls of sin before taking flight for a better world, was fain to tell them "there was no hurry, because the eclipse had been put off a fortnight on account of the number of penitents"!

[Illustration: FIG. 76.--Eclipse of the Moon at Laos (February 27, 1877).]

These fears and terrors are still extant among ignorant peoples. In the night of February 27, 1877, an eclipse of the Moon produced an indescribable panic among the inhabitants of Laos (Indo-China). In order to frighten off the Black Dragon, the natives fired shots at the half-devoured orb, accompanying their volley with the most appalling yells. Dr. Harmand has memorialized the scene in the lively sketch given on p. 269.

During the solar eclipse of March 15, 1877, an analogous scene occurred among the Turks, who for the moment forgot their preparations for war with Russia, in order to shoot at the Sun, and deliver him from the toils of the Dragon.

The lunar eclipse of December 16, 1880, was not unnoticed at Tackhent (Russian Turkestan), where it was received with a terrific din of saucepans, samovars and various implements struck together again and again by willing hands that sought to deliver the Moon from the demon Tchaitan who was devouring her.

In China, eclipses are the object of imposing ceremonies, whose object is to reestablish the regularity of the celestial motions. Since the Emperor is regarded as the Son of Heaven, his government must in some sort be a reflection of the immutable order of the sidereal harmonies. As eclipses were regarded by astrologers as disturbances of the divine order, their appearance indicates some irregularity in the government of the Celestial Empire. Accordingly, they are received with all kinds of expiatory ceremonies prescribed thousands of years ago, and still in force to-day.

In the twentieth century, as in the nineteenth, the eighteenth, or in ancient epochs, the same awe and terror operates upon the ignorant populations who abound upon the surface of our planet.

To return to astronomical realities.

We said above that these phenomena were produced when the Full Moon and the New Moon reached the line of intersection, known as the line of nodes, when the plane of the lunar orbit cuts the plane of the ecliptic. As this line turns and comes back in the same direction relatively to the Sun at the end of eighteen years, eleven days, we have only to register the eclipses observed during this period in order to know all that will occur in the future, and to find such as happened in the past. This period was known to the Greeks under the name of the Metonic Cycle, and the Chaldeans employed it three thousand years ago under the name of Saros.

On examining this cycle, composed of 223 lunations, we see that there can not be more than seven eclipses in one year, nor less than two. When there are only two, they are eclipses of the Sun.

The totality of a solar eclipse can not last more than seven minutes, fifty-eight seconds at the equator, and six minutes, ten seconds in the latitude of Paris. The Moon, on the contrary, may be entirely eclipsed for nearly two hours.

Eclipses of the Sun are very rare for a definite spot. Thus not one occurred for Paris during the whole of the nineteenth century, the last which happened exactly above the capital of France having been on May 22, 1724. I have calculated all those for the twentieth century, and find that two will take place close to Paris, on April 17, 1912, at eighteen minutes past noon (total for Choisy-le-Roi, Longjumeau, and Dourdan, but very brief: seven seconds), and August 11, 1999, at 10.28 A.M. (total for Beauvais, Compi 鐇 ne, Amiens, St. Quentin, fairly long: two minutes, seventeen seconds). Paris itself will not be favored before August 12, 2026. In order to witness the phenomenon, one must go and look for it. This the author did on May 28, 1900, in Spain.

The progress of the lunar shadow upon the surface of the Earth is traced beforehand on maps that serve to show the favored countries for which our satellite will dispense her ephemeral night. The above figure shows the trajectory of the total phase of the 1900 eclipse in Portugal, Spain, Algeria, and Tunis.

[Illustration: FIG. 77.--The path of the Eclipse of May 28, 1900.]

The immutable splendor of the celestial motions had never struck the author so impressively as during the observation of this grandiose phenomenon. With the absolute precision of astronomical calculations, our satellite, gravitating round the Earth, arrived upon the theoretical line drawn from the orb of day to our planet, and interposed itself gradually, slowly, and exactly, in front of it. The eclipse was total, and occurred at the moment predicted by calculation. Then the obscure globe of the Moon pursued its regular course, discovered the radiant orb behind, and gradually and slowly completed its transit in front of him. Here, to all observers, was a double philosophical lesson, a twofold impression: that of the greatness, the omnipotence of the inexorable forces that govern the universe, and that of the inexorable valor of man, of this thinking atom straying upon another atom, who by the travail of his feeble intelligence has arrived at the knowledge of the laws by which he, like the rest of the world, is borne away through space, through time, and through eternity.

The line of centrality passed through Elche, a picturesque city of 30,000 inhabitants, not far from Alicante, and we had chosen this for our station on account of the probability of fine weather.

From the terrace of the country house of the hospitable Mayor, a farm transformed into an observatory by our learned friend, Count de la Baume Pluvinel, there were no obstacles between ourselves and any part of the sky or landscape. The whole horizon lay before us. In front was a town of Arab aspect framed in a lovely oasis of palm-trees; a little farther off, the blue sea beyond the shores of Alicante and Murcia: on the other side a belt of low mountains, and near us fields and gardens. A Company of the Civic Guard kept order, and prevented the entrance of too many curious visitors, of whom over ten thousand had arrived.

At the moment when the first contact of the lunar disk with the solar disk was observed in the telescope, we fired a gun, in order to announce the precise commencement of the occultation to the 40,000 persons who were awaiting the phenomenon, and to discover what difference would exist between this telescopic observation and those made with the unaided eyes (protected simply by a bit of smoked glass) of so many improvised spectators. This had already been done by Arago at Perpignan in 1842. The verification was almost immediate for the majority of eyes, and may be estimated at eight or ten seconds. So that the commencement of the eclipse was confirmed almost as promptly for the eye as with the astronomical instruments.

The sky was splendidly clear; no cloud, no mist, deep blue; blazing Sun. The first period of the eclipse showed nothing particular. It is only from the moment when more than half the solar disk is covered by the lunar disk that the phenomenon is imposing in its grandeur. At this phase, I called the attention of the people standing in the court to the visibility of the stars, and indicating the place of Venus in the sky asked if any with long sight could perceive her. Eight at once responded in the affirmative. It should be said that the planet was at that time at its period of maximum brilliancy, when for observers blessed with good sight, it is always visible to the unaided eye.

When some three-quarters of the Sun were eclipsed, the pigeons which had flown back to the farm huddled into a corner, and made no further movement. They told me that evening that the fowls had done the same a little later, returning to the hen-house as though it had been night, and that the small children (who were very numerous at Elche, where the population is certainly not diminishing) left off their games, and came back to their mothers' skirts. The birds flew anxiously to their nests. The ants in one garden were excessively agitated, no doubt disconcerted in their strategics. The bats came out.

A few days before the eclipse I had prepared the inhabitants of this part of Spain for the observation of the phenomenon by the following description, which sums up the previous accounts of the astronomers:

"The spectacle of a total eclipse of the Sun is one of the most magnificent and imposing that it is possible to see in nature. At the exact moment indicated by calculation, the Moon arrives in front of the Sun, eats into it gradually, and at last entirely covers it. The light of the day lessens and is transformed. A sense of oppression is felt by all nature, the birds are hushed, the dog takes refuge with his master, the chickens hide beneath their mother's wing, the wind drops, the temperature falls, an appalling stillness is everywhere perceptible, as though the universe were on the verge of some imminent catastrophe. Men's faces assume a cadaverous hue similar to that given at night by the flame of spirits of wine and salt, a livid funereal light, the sinister illumination of the world's last hour.

"At the moment when the last line of the solar crescent disappears, we see, instead of the Sun, a black disk surrounded with a splendid luminous aureole shooting immense jets into space, with roseate flames burning at the base.

"A sudden night has fallen on us, a weird, wan night in which the brightest of the stars are visible in the Heavens. The spectacle is splendid, grandiose, solemn, and sublime."

This impression was actually felt by us all, as may be seen from the following notes, written in my schedule of observation during the event, or immediately after:

"3.50 P.M. Light very weak, sky leaden gray, mountains standing out with remarkable clearness from the horizon, and seeming to approach us.

"3.55 P.M. Fall of temperature very apparent. Cold wind blowing through the atmosphere.

"3.56 P.M. Profound silence through nature, which seems to participate in the celestial phenomenon. Silence in all the groups.

"3.57 P.M. Light considerably diminished, becoming wan, strange, and sinister. Landscape leaden gray, sea looks black. This diminution of light is not that of every day after the sunset. There is, as it were, a tint of sadness spread over the whole of nature. One becomes accustomed to it, and yet while we know that the occultation of the Sun by the Moon is a natural phenomenon, we can not escape a certain sense of uneasiness. The approach of some extraordinary spectacle is imminent."

At this point we examined the effects of the solar light upon the seven colors of the spectrum. In order to determine as accurately as possible the tonality of the light of the eclipse, I had prepared seven great sheets, each painted boldly in the colors of the spectrum, violet, indigo, blue, green, yellow, orange, red; and a similar series in pieces of silk. These colors were laid at our feet upon the terrace where my wife, as well as Countess de la Baume, were watching with me. We then saw the first four disappear successively and entirely and turn black in a few seconds, in the following order: violet, indigo, blue, green. The three other colors were considerably attenuated by the darkness, but remained visible.

It should be noted that in the normal order of things--that is, every evening-- the contrary appears; violet remains visible after the red.

This experiment shows that the last light emitted by the eclipsed Sun belongs to the least refrangible rays, to the greatest wave-lengths, to the slowest vibrations, to the yellow and red rays. Such therefore is the predominating color of the solar atmosphere.

This experiment completed, we turn back to the Sun. Magical and splendid spectacle! Totality has commenced, the Sun has disappeared, the black disk of the Moon covers it entirely, leaving all round it a magnificent corona of dazzling light. One would suppose it to be an annular eclipse, with the difference that this can be observed with the naked eye, without fatigue to the retina, and drawn quietly.

This luminous coronal atmosphere entirely surrounds the solar disk, at a pretty equal depth, equivalent to about the third of half the solar diameter. It may be regarded as the Sun's atmosphere.

Beyond this corona is an aureole, of vaster glory but less luminous, which sends out long plumes, principally in the direction of the equatorial zone of the Sun, and of the belt of activity of the spots and prominences.

At the summit of the disk it is conical in shape. Below it is double, and its right-hand portion ends in a point, not far from Mercury, which shines like a dazzling star of first magnitude, and seems placed there expressly to give us the extent and direction of the solar aureole.

I draw these various aspects (which, moreover, change with the movement of the Moon), and what strikes me most is the distinction in light between this aureole and the coronal atmosphere; the latter appears to be a brilliant silvery white, the former is grayer and certainly less dense.

My impression is that there are two solar envelopes of entirely different nature, the corona belonging to the globe of the Sun, and forming its atmosphere properly so-called, very luminous; the aureole formed of particles that circulate independently round it, probably arising from eruptions, their form as a whole being possibly due to electric or magnetic forces, counterbalanced by resistances of various natures. In our own atmosphere the volcanic eruptions are distinct from the aerial envelope.

The general configuration of this external halo, spreading more particularly in the equatorial zone, is sufficiently like that of the eclipse of 1889, published in my Popular Astronomy, which also corresponded with a minimum of solar energy. The year 1900 is in fact close upon the minimum of the eleven-year period. This equatorial form is, moreover, what all the astronomers were expecting.

[Illustration: FIG. 78.--Total eclipse of the Sun, May 28, 1900, as observed from Elche (Spain).]

There can no longer be the slightest doubt that the solar envelope varies with the activity of the Sun....

"But the total eclipse lasted a much shorter time than I have taken to write these lines. The seventy-nine seconds of totality are over. A dazzling light bursts from the Sun, and tells that the Moon pursuing its orbit has left it. The splendid sight is over. It has gone like a shadow.

"Already over! It is almost a disillusion. Nothing beautiful lasts in this world. Too sad! If only the celestial spectacle could have lasted two, three, or four minutes! It was too short....

"Alas! we are forced to take things as they are.

"The surprise, the oppression, the terror of some, the universal silence are over. The Sun reappears in his splendor, and the life of nature resumes its momentarily suspended course.

"While I was making my drawing, M. l'Abb?Moreux, my colleague from the Astronomical Society of France, who accompanied me to Spain for this observation, was taking one of his own, without any reciprocal communication. These two sketches are alike, and confirmatory.

"The differential thermometers that I exposed to the Sun, hanging freely, and protected from reflection from the ground, were read every five minutes. The black thermometer went down from 33.1?to 20.7? that is 12.4? the white from 29?to 20.2?-that is, 8.8? The temperature in the shade only varied three degrees.

"The light received during totality was due: first, to the luminous envelope of the Sun; second, to that of the terrestrial atmosphere, illuminated at forty kilometers (twenty-five miles) on the one side and the other of the line of centrality. It appeared to be inferior to that of the Full Moon, on account of the almost sudden transition. But, in reality, it was more intense, for only first-magnitude stars were visible in the sky, whereas on a night of full moon, stars of second, and even of third magnitude are visible. We recognized, among others, Venus, Mercury, Sirius, Procyon, Capella, Rigel, Betelgeuse."

* * * * *

From these notes, taken on the spot, it is evident that the contemplation of a total eclipse of the Sun is one of the most marvelous spectacles that can be admired upon our planet.

Some persons assured me that they saw the shadow of the Moon flying rapidly over the landscape. My attention was otherwise occupied, and I was unable to verify this interesting observation. The shadow of the Moon in effect took only eleven minutes (3.47 P.M. to 3.58 P.M.) to traverse the Iberian Peninsula from Porto to Alicante, i.e., a distance of 766 kilometers (475 miles). It must therefore have passed over the ground at a velocity of sixty-nine kilometers per minute, or 1,150 meters per second, a speed higher than that of a bullet. It can easily be watched from afar, on the mountains.

Some weeks previous to this fine eclipse, when I informed the Spaniards of the belt along which it could be observed, I had invited them to note all the interesting phenomena they might witness, including the effects produced by the eclipse upon animals. Birds returned hurriedly to their nests, swallows lost themselves, sheep huddled into compact packs, partridges were hypnotized, frogs croaked as if it were night, fowls took refuge in the hen-house, and cocks crowed, bats came out, and were surprised by the sun, chicks gathered under their mothers' wing, cage-birds ceased their songs, some dogs howled, others crept shivering to their masters' feet, ants returned to the antheap, grasshoppers chirped as at sunset, pigeons sank to the ground, a swarm of bees went silently back to their hive, and so on.

These creatures behaved as though the night had come, but there were also signs of fear, surprise, even of terror, differing only "in degree" from those manifested during the grandiose phenomenon of a total eclipse by human beings unenlightened by a scientific education.

At Madrid the eclipse was only partial. The young King of Spain, Alfonso XIII, took care to photograph it, and I offer the photograph to my readers (Fig. 79), as this amiable sovereign did me the honor to give it me a few days after the eclipse.

[Illustration: FIG. 79.--The Eclipse of May 28, 1900, as photographed by King Alfonso XIII, at Madrid.]

The technical results of these observations of solar eclipses relate more especially to the elucidation of the grand problem of the physical constitution of the Sun. We alluded to them in the chapter devoted to this orb. The last great total eclipses have been of immense value to science.

The eclipses of the Moon are less important, less interesting, than the eclipses of the Sun. Yet their aspect must not be neglected on this account, and it may be said to vary for each eclipse.

Generally speaking, our satellite does not disappear entirely in the Earth's cone of shadow; the solar rays are refracted round our globe by our atmosphere, and curving inward, illumine the lunar globe with a rosy tint that reminds one of the sunset. Sometimes, indeed, this refraction does not occur, owing doubtless to lack of transparency in the atmosphere, and the Moon becomes invisible. This happened recently, on April 11, 1903.

For any spot, eclipses of the Moon are incomparably more frequent than eclipses of the Sun, because the cone of lunar shadow that produces the solar eclipses is not very broad at its contact with the surface of the globe (10, 20, 30, 50, 100 kilometers, according to the distance of the Moon), whereas all the countries of the Earth for which the Moon is above the horizon at the hour of the lunar eclipse are able to see it. It is at all times a remarkable spectacle that uplifts our thoughts to the Heavens, and I strongly advise my readers on no account to forego it.

CHAPTER XI

ON METHODS

HOW CELESTIAL DISTANCES ARE DETERMINED, AND HOW THE SUN IS WEIGHED

I will not do my readers the injustice to suppose that they will be alarmed at the title of this Lesson, and that they do not employ some "method" in their own lives. I even assume that if they have been good enough to take me on faith when I have spoken of the distances of the Sun and Moon, and Stars, or of the weight of bodies at the surface of Mars, they retain some curiosity as to how the astronomers solve these problems. Hence it will be as interesting as it is useful to complete the preceding statements by a brief summary of the methods employed for acquiring these bold conclusions.

The Sun seems to touch the Earth when it disappears in the purple mists of twilight: an immense abyss separates us from it. The stars go hand in hand down the constelled sky; and yet one can not think of their inconceivable distance without a shiver.

Our neighbor, Moon, floats in space, a stone's throw from us: but without calculation we should never know the distance, which remains an impassable desert to us.

The best educated persons sometimes find it difficult to admit that these distances of Sun and Moon are better determined and more precise than those of certain points on our minute planet. Hence, it is of particular moment for us to give an exact account of the means employed in determining them.

The calculation of these distances is made by "triangulation." This process is the same that surveyors use in the measurement of terrestrial distances. There is nothing very alarming about it. If the word repels us a little at first, it is from its appearance only.

When the distance of an object is unknown, the only means of expressing its apparent size is by measurement of the angle which it subtends before our eyes.

We all know that an object appears smaller, in proposition with its distance from us. This diminution is not a matter of chance. It is geometric, and proportional to the distance. Every object removed to a distance of 57 times its diameter measures an angle of 1 degree, whatever its real dimensions. Thus a sphere 1 meter in diameter measures exactly 1 degree, if we see it at a distance of 57 meters. A statue measuring 1.80 meters (about 5 ft. 8 in.) will be equal to an angle of 1 degree, if distant 57 times its height, that is to say, at 102.60 meters. A sheet of paper, size 1 decimeter, seen at 5.70 meters, represents the same magnitude.

In length, a degree is the 57th part of the radius of a circle, i.e., from the circumference to the center.

The measurement of an angle is expressed in parts of the circumference. Now, what is an angle of a degree? It is the 360th part of any circumference. On a table 3.60 meters round, an angle of one degree is a centimeter, seen from the center of the table. Trace on a sheet of paper a circle 0.360 meters round--an angle of 1 degree is a millimeter.

[Illustration: FIG. 80.--Measurement of Angles.]

If the circumference of a circus measuring 180 meters be divided into 360 places, each measuring 0.50 meters in width, then when the circus is full a person placed at the center will see each spectator occupying an angle of 1 degree. The angle does not alter with the distance, and whether it be measured at 1 meter, 10 meters, 100 kilometers, or in the infinite spaces of Heaven, it is always the same angle. Whether a degree be represented by a meter or a kilometer, it always remains a degree. As angles measuring less than a degree often have to be calculated, this angle has been subdivided into 60 parts, to which the name of minutes has been given, and each minute into 60 parts or seconds. Written short, the degree is indicated by a little zero (? placed above the figure; the minute by an apostrophe ('), and the second by two ("). These minutes and seconds of arc have no relation with the same terms as employed for the division of the duration of time. These latter ought never to be written with the signs of abbreviation just indicated, though journalists nowadays set a somewhat pedantic example, by writing, e.g., for an automobile race, 4h. 18' 30", instead of 4h. 18m. 30s.

This makes clear the distinction between the relative measure of an angle and the absolute measures, such, for instance, as the meter. Thus, a degree may be measured on this page, while a second (the 3,600th part of a degree) measured in the sky may correspond to millions of kilometers.

Now the measure of the Moon's diameter gives us an angle of a little more than half a degree. If it were exactly half a degree, we should know by that that it was 114 times the breadth of its disk away from us. But it is a little less, since we have more than half a degree (31'), and the geometric ratio tells us that the distance of our satellite is 110 times its diameter.

Hence we have very simply obtained a first idea of the distance of the Moon by the measure of its diameter. Nothing could be simpler than this method. The first step is made. Let us continue.

This approximation tells us nothing as yet of the real distance of the orb of night. In order to know this distance in miles, we need to know the width in miles of the lunar disk.

[Illustration: FIG. 81.--Division of the Circumference into 360 degrees.]

This problem has been solved, as follows:

Two observers go as far as possible from each other, and observe the Moon simultaneously, from two stations situated on the same meridian, but having a wide difference of latitude. The distance that separates the two points of observation forms the base of a triangle, of which the two long sides come together on the Moon.

[Illustration: FIG. 82.--Measurement of the distance of the Moon.]

It is by this proceeding that the distance of our satellite was finally established, in 1751 and 1752, by two French astronomers, Lalande and Lacaille; the former observing at Berlin, the latter at the Cape of Good Hope. The result of their combined observations showed that the angle formed at the center of the lunar disk by the half-diameter of the Earth is 57 minutes of arc (a little less than a degree). This is known as the parallax of the Moon.

Here is a more or less alarming word; yet it is one that we can not dispense with in discussing the distance of the stars. This astronomical term will soon become familiar in the course of the present lesson, where it will frequently recur, and always in connection with the measurement of celestial distances. "Do not let us fear," wrote Lalande in his Astronomie des Dames, "do not let us fear to use the term parallax, despite its scientific aspect; it is convenient, and this term explains a very simple and very familiar effect."

"If one is at the play," he continues, "behind a woman whose hat is too large, and prevents one from seeing the stage [written a hundred years ago!], one leans to the left or right, one rises or stoops: all this is a parallax, a diversity of aspect, in virtue of which the hat appears to correspond with another part of the theater from that in which are the actors." "It is thus," he adds, "that there may be an eclipse of the Sun in Africa and none for us, and that we see the Sun perfectly, because we are high enough to prevent the Moon's hiding it from us."

See how simple it is. This parallax of 57 minutes proves that the Earth is removed from the Moon at a distance of about 60 times its half-diameter (precisely, 60.27). From this to the distance of the Moon in kilometers is only a step, because it suffices to multiply the half-diameter of the Earth, which is 6,371 kilometers (3,950 miles) by this number. The distance of our satellite, accordingly, is 6,371 kilometers, multiplied by 60.27--that is, 384,000 kilometers (238,000 miles). The parallax of the Moon not only tells us definitely the distance of our planet, but also permits us to calculate its real volume by the measure of its apparent volume. As the diameter of the Moon seen from the Earth subtends an angle of 31', while that of the Earth seen from the Moon is 114', the real diameter of the orb of night must be to that of the terrestrial globe in the relation of 273 to 1,000. That is a little more than a quarter, or 3,480 kilometers (2,157 miles), the diameter of our planet being 12,742 kilometers (7,900 miles).

This distance, calculated thus by geometry, is positively determined with greater precision than that employed in the ordinary measurements of terrestrial distances, such as the length of a road, or of a railway. This statement may seem to be a romance to many, but it is undeniable that the distance separating the Earth from the Moon is measured with greater care than, for instance, the length of the road from Paris to Marseilles, or the weight of a pound of sugar at the grocer's. (And we may add without comment, that the astronomers are incomparably more conscientious in their measurements than the most scrupulous shop-keepers.)

Had we conveyed ourselves to the Moon in order to determine its distance and its diameter directly, we should have arrived at no greater precision, and we should, moreover, have had to plan out a journey which in itself is the most insurmountable of all the problems.

The Moon is at the frontier of our little terrestrial province: one might say that it traces the limits of our domain in space. And yet, a distance of 384,000 kilometers (238,000 miles) separates the planet from the satellite. This space is insignificant in the immeasurable distances of Heaven: for the Saturnians (if such exist!) the Earth and the Moon are confounded in one tiny star; but for the inhabitants of our globe, the distance is beyond all to which we are accustomed. Let us try, however, to span it in thought.

A cannon-ball at constant speed of 500 meters (547 yards) per second would travel 8 days, 5 hours to reach the Moon. A train started at a speed of one kilometer per minute, would arrive at the end of an uninterrupted journey in 384,000 minutes, or 6,400 hours, or 266 days, 16 hours. And in less than the time it takes to write the name of the Queen of Night, a telegraphic message would convey our news to the Moon in one and a quarter seconds.

Long-distance travelers who have been round the world some dozen times have journeyed a greater distance.

The other stars (beginning with the Sun) are incomparably farther from us. Yet it has been found possible to determine their distances, and the same method has been employed.

But it will at once be seen that different measures are required in calculating the distance of the Sun, 388 times farther from us than the Moon, for from here to the orb of day is 12,000 times the breadth of our planet. Here we must not think of erecting a triangle with the diameter of the Earth for its base: the two ideal lines drawn from the extremities of this diameter would come together between the Earth and the Sun; there would be no triangle, and the measurement would be absurd.

In order to measure the distance which separates the Earth from the Sun, we have recourse to the fine planet Venus, whose orbit is situated inside the terrestrial orbit. Owing to the combination of the Earth's motion with that of the Star of the Morning and Evening, the capricious Venus passes in front of the Sun at the curious intervals of 8 years, 113-1/2 years less 8 years, 8 years, 113-1/2 years plus 8 years.

Thus there was a transit in June, 1761, then another 8 years after, in June,

1769. The next occurred 113-1/2 years less 8 years, i.e., 105-1/2 years after the preceding, in December, 1874; the next in December, 1882. The next will be in June, 2004, and June, 2012. At these eagerly anticipated epochs, astronomers watch the transit of Venus across the Sun at two terrestrial stations as far as possible removed from each other, marking the two points at which the planet, seen from their respective stations, appears to be projected at the same moment on the solar disk. This measure gives the width of an angle formed by two lines, which starting from two diametrically opposite points of the Earth, cross upon Venus, and form an identical angle upon the Sun. Venus is thus at the apex of two equal triangles, the bases of which rest, respectively, upon the Earth and on the Sun. The measurement of this angle gives what is called the parallax of the Sun--that is, the angular dimension at which the Earth would be seen at the distance of the Sun.

[Illustration: FIG. 83.--Measurement of the distance of the Sun.]

Thus, it has been found that the half-diameter of the Earth viewed from the Sun measures 8.82". Now, we know that an object presenting an angle of one degree is at a distance of 57 times its length.

The same object, if it subtends an angle of a minute, or the sixtieth part of a degree, indicates by the measurement of its angle that it is 60 times more distant, i.e., 3,438 times.

Finally, an object that measures one second, or the sixtieth part of a minute, is at a distance of 206,265 times its length.

Hence we find that the Earth is at a distance from the Sun of 206,265/8.82--that is, 23,386 times its half-diameter, that is, 149,000,000 kilometers (93,000,000 miles). This measurement again is as precise and certain as that of the Moon.

I hope my readers will easily grasp this simple method of triangulation, the result of which indicates to us with absolute certainty the distance of the two great celestial torches to which we owe the radiant light of day and the gentle illumination of our nights.

The distance of the Sun has, moreover, been confirmed by other means, whose results agree perfectly with the preceding. The two principal are based on the velocity of light. The propagation of light is not instantaneous, and notwithstanding the extreme rapidity of its movements, a certain time is required for its transmission from one point to another. On the Earth, this velocity has been measured as 300,000 kilometers (186,000 miles) per second. To come from Jupiter to the Earth, it requires thirty to forty minutes, according to the distance of the planet. Now, in examining the eclipses of Jupiter's satellites, it has been discovered that there is a difference of 16 minutes, 34 seconds in the moment of their occurrence, according as Jupiter is on one side or on the other of the Sun, relatively to the Earth, at the minimum and maximum distance. If the light takes 16 minutes, 34 seconds to traverse the terrestrial orbit, it must take less than that time, or 8 minutes, 17 seconds, to come to us from the Sun, which is situated at the center. Knowing the velocity of light, the distance of the Sun is easily found by multiplying 300,000 by 8 minutes, 17 seconds, or 497 seconds, which gives about 149,000,000 kilometers (93,000,000 miles).

Another method founded upon the velocity of light again gives a confirmatory result. A familiar example will explain it: Let us imagine ourselves exposed to a vertical rain; the degree of inclination of our umbrella will depend on the relation between our speed and that of the drops of rain. The more quickly we run, the more we need to dip our umbrella in order not to meet the drops of water. Now the same thing occurs for light. The stars, disseminated in space, shed floods of light upon the Heavens. If the Earth were motionless, the luminous rays would reach us directly. But our planet is spinning, racing, with the utmost speed, and in our astronomical observations we are forced to follow its movements, and to incline our telescopes in the direction of its advance. This phenomenon, known under the name of aberration of light, is the result of the combined effects of the velocity of light and of the Earth's motion. It shows that the speed of our globe is equivalent to 1/10000 that of light, i.e., = about 30 kilometers (19 miles) per second. Our planet accordingly accomplishes her revolution round the Sun along an orbit which she traverses at a speed of 30 kilometers (better 29-1/2) per second, or 1,770 kilometers per minute, or 106,000 kilometers per hour, or 2,592,000 kilometers per day, or 946,080,000 kilometers (586,569,600 miles) in the year. This is the length of the elliptical path described by the Earth in her annual translation.

The length of orbit being thus discovered, one can calculate its diameter, the half of which is exactly the distance of the Sun.

We may cite one last method, whose data, based upon attraction, are provided by the motions of our satellite. The Moon is a little disturbed in the regularity of her course round the Earth by the influence of the powerful Sun. As the attraction varies inversely with the square of the distance, the distance may be determined by analyzing the effect it has upon the Moon.

Other means, on which we will not enlarge in this summary of the methods employed for determinations, confirm the precisions of these measurements with certainty. Our readers must forgive us for dwelling at some length upon the distance of the orb of day, since this measurement is of the highest importance; it serves as the base for the valuation of all stellar distances, and may be considered as the meter of the universe.

This radiant Sun to which we owe so much is therefore enthroned in space at a distance of 149,000,000 kilometers (93,000,000 miles) from here. Its vast brazier must indeed be powerful for its influence to be exerted upon us to such a manifest extent, it being the very condition of our existence, and reaching out as far as Neptune, thirty times more remote than ourselves from the solar focus.

It is on account of its great distance that the Sun appears to us no larger than the Moon, which is only 384,000 kilometers (238,000 miles) from here, and is itself illuminated by the brilliancy of this splendid orb.

No terrestrial distance admits of our conceiving of this distance. Yet, if we associate the idea of space with the idea of time, as we have already done for the Moon, we may attempt to picture this abyss. The train cited just now would, if started at a speed of a kilometer a minute, arrive at the Sun after an uninterrupted course of 283 years, and taking as long to return to the Earth the total would be 566 years. Fourteen generations of stokers would be employed on this celestial excursion before the bold travelers could bring back news of the expedition to us.

Sound is transmitted through the air at a velocity of 340 meters (1,115 feet) per second. If our atmosphere reached to the Sun, the noise of an explosion sufficiently formidable to be heard here would only reach us at the end of 13 years, 9 months. But the more rapid carriers, such as the telegraph, would leap across to the orb of day in 8 minutes, 17 seconds.

Our imagination is confounded before this gulf of 93,000,000 miles, across which we see our dazzling Sun, whose burning rays fly rapidly through space in order to reach us.

* * * * *

And now let us see how the distances of the planets were determined.

We will leave aside the method of which we have been speaking; that now to be employed is quite different, but equally precise in its results.

It is obvious that the revolution of a planet round the Sun will be longer in proportion as the distance is greater, and the orbit that has to be traveled vaster. This is simple. But the most curious thing is that there is a geometric proportion in the relations between the duration of the revolutions of the planets and their distances. This proportion was discovered by Kepler, after thirty years of research, and embodied in the following formula:

"The squares of the times of revolution of the planets round the Sun (the periodic times) are proportional to the cubes of their mean distances from the Sun."

This is enough to alarm the boldest reader. And yet, if we unravel this somewhat incomprehensible phrase, we are struck with its simplicity.

What is a square? We all know this much; it is taught to children of ten years old. But lest it has slipped your memory: a square is simply a number multiplied by itself.

Thus: 2 ?2 = 4; 4 is the square of 2.

Four times 4 is 16; 16 is the square of 4.

And so on, indefinitely.

Now, what is a cube? It is no more difficult. It is a number multiplied twice by itself.

For instance: 2 multiplied by 2 and again by 2 equals 8. So 8 is the cube of 2. 3 ?3 ?3 = 27; 27 is the cube of 3, and so on.

Now let us take an example that will show the simplicity and precision of the formula enunciated above. Let us choose a planet, no matter which. Say, Jupiter, the giant of the worlds. He is the Lord of our planetary group. This colossal star is five times (precisely, 5.2) as far from us as the Sun.

Multiply this number twice by itself 5.2 ?5.2 ?5.2 = 140.

On the other hand, the revolution of Jupiter takes almost twelve years (11.85). This number multiplied by itself also equals 140. The square of the number 11.85 is equal to the cube of the number 5.2. This very simple law regulates all the heavenly bodies.

Thus, to find the distance of a planet, it is sufficient to observe the time of its revolution, then to discover the square of the given number by multiplying it into itself. The result of the operation gives simultaneously the cube of the number that represents the distance.

To express this distance in kilometers (or miles), it is sufficient to multiply it by 149,000,000 (in miles 93,000,000), the key to the system of the world.

Nothing, then, could be less complicated than the definition of these methods. A few moments of attention reveal to us in their majestic simplicity the immutable laws that preside over the immense harmony of the Heavens.

* * * * *

But we must not confine ourselves to our own solar province. We have yet to speak of the stars that reign in infinite space far beyond our radiant Sun.

Strange and audacious as it may appear, the human mind is able to cross these heights, to rise on the wings of genius to these distant suns, and to plumb the depths of the abyss that separates us from these celestial kingdoms.

Here, we return to our first method, that of triangulation. And the distance that separates us from the Sun must serve in calculating the distances of the stars.

The Earth, spinning round the Sun at a distance of 149,000,000 kilometers (93,000,000 miles), describes a circumference, or rather an ellipse, of 936,000,000 kilometers (580,320,000 miles), which it travels over in a year. The distance of any point of the terrestrial orbit from the diametrically opposite point which it passes six months later is 298,000,000 kilometers (184,760,000 miles), i.e., the diameter of this orbit. This immense distance (in comparison with those with which we are familiar) serves as the base of a triangle of which the apex is a star.

The difficulty in exact measurements of the distance of a star consists in observing the little luminous point persistently for a whole year, to see if this star is stationary, or if it describes a minute ellipse reproducing in perspective the annual revolution of the Earth.

If it remains fixed, it is lost in such depths of space that it is impossible to gage the distance, and our 298,000,000 kilometers have no meaning in view of such an abyss. If, on the contrary, it is displaced, it will in the year describe a minute ellipse, which is only the reflection, the perspective in miniature, of the revolution of our planet round the Sun.

The annual parallax of a star is the angle under which one would see the radius, or half-diameter, of the terrestrial orbit from it. This radius of 149,000,000 kilometers (93,000,000 miles) is indeed, as previously observed, the unit, the meter of celestial measures. The angle is of course smaller in proportion as the star is more distant, and the apparent motion of the star diminishes in the same proportion. But the stars are all so distant that their annual displacement of perspective is almost imperceptible, and very exact instruments are required for its detection.

[Illustration: FIG. 84.--Small apparent ellipses described by the stars as a result of the annual displacement of the Earth.]

The researches of the astronomers have proved that there is not one star for which the parallax is equal to that of another. The minuteness of this angle, and the extraordinary difficulties experienced in measuring the distance of the stars, will be appreciated from the fact that the value of a second is so small that the displacement of any star corresponding with it could be covered by a spider's thread.

A second of arc corresponds to the size of an object at a distance of 206,265 times its diameter; to a millimeter seen at 206 meters' distance; to a hair, 1/10 of a millimeter in thickness, at 20 meters' distance (more invisible to the naked eye). And yet this value is in excess of those actually obtained. In fact:--the apparent displacement of the nearest star is calculated at 75/100 of a second (0.75"), i.e., from this star, [alpha] of Centaur, the half-diameter of the terrestrial orbit is reduced to this infinitesimal dimension. Now in order that the length of any straight line seen from the front be reduced until it appear to subtend no more than an angle of 0.75", it must be removed to a distance 275,000 times its length. As the radius of the terrestrial orbit is 149,000,000 kilometers (93,000,000 miles), the distance which separates [alpha] of Centaur from our world must therefore = 41,000,000,000,000 kilometers (25,000,000,000,000 miles). And that is the nearest star. We saw in Chapter II that it shines in the southern hemisphere. The next, and one that can be seen in our latitudes, is 61 of Cygnus, which floats in the Heavens 68,000,000,000,000 kilometers (42,000,000,000,000 miles) from here. This little star, of fifth magnitude, was the first of which the distance was determined (by Bessel, 1837-1840).

All the rest are much more remote, and the procession is extended to infinity.

We can not conceive directly of such distances, and in order to imagine them we must again measure space by time.

In order to cover the distance that separates us from our neighbor, [alpha] of Centaur, light, the most rapid of all couriers, takes 4 years, 128 days. If we would follow it, we must not jump from start to finish, for that would not give us the faintest idea of the distance: we must take the trouble to think out the direct advance of the ray of light, and associate ourselves with its progress. We must see it traverse 300,000 kilometers (186,000 miles) during the first second of the journey; then 300,000 more in the second, which makes 600,000 kilometers; then once more 300,000 kilometers during the third, and so on without stopping for four years and four months. If we take this trouble we may realize the value of the figure; otherwise, as this number surpasses all that we are in the habit of realizing, it will have no significance for us, and will be a dead letter.

If some appalling explosion occurred in this star, and the sound in its flight of 340 meters (1,115 feet) per second were able to cross the void that separates us from it, the noise of this explosion would only reach us in 3,000,000 years.

A train started at a speed of 106 kilometers (65 miles) per hour would have to run for 46,000,000 years, in order to reach this star, our neighbor in the celestial kingdom.

The distance of some thirty of the stars has been determined, but the results are dubious.

The dazzling Sirius reigns 92,000,000,000,000 kilometers (57,000,000,000,000 miles), the pale Vega at 204,000,000,000,000. Each of these magnificent stars must be a huge sun to burn at such a distance with such luminosity. Some are millions of times larger than the Earth. Most of them are more voluminous than our Sun. On all sides they scintillate at inaccessible distances, and their light strays a long while in space before it encounters the Earth. The luminous ray that we receive to-day from some pale star hardly perceptible to our eyes--so enormous is its distance--may perhaps bring us the last emanation of a sun that expired thousands of years ago.

* * * * *

If these methods have been clear to my readers, they may also be interested perhaps in knowing the means employed in weighing the worlds. The process is as simple and as clear as those of which we have been speaking.

Weighing the stars! Such a pretension seems Utopian, and one asks oneself curiously what sort of balance the astronomers must have adopted in order to calculate the weight of Sun, Moon, planets or stars.

Here, figures replace weights. Ladies proverbially dislike figures: yet it would be easier for some society dame to weigh the Sun at the point of her pen, by writing down a few columns of figures with a little care, than to weigh a 12 kilogram case of fruit, or a dress-basket of 35 kilos, by direct methods.

Weighing the Sun is an amusement like any other, and a change of occupation.

If the Moon were not attracted by the Earth, she would glide through the Heavens along an indefinite straight line, escaping at the tangent. But in virtue of the attraction that governs the movements of all the Heavenly bodies, our satellite at a distance of 60 times the terrestrial half-diameter revolves round us in 27 days, 7 hours, 43 minutes, 11-1/2 seconds, continually leaving the straight line to approach the Earth, and describing an almost circular orbit in space. If at any moment we trace an arc of the lunar orbit, and if a tangent is taken to this arc, the deviation from the straight line caused by the attraction of our planet is found to be 1-1/3 millimeter per second.

This is the quantity by which the Moon drops toward us in each second, during which she accomplishes 1,017 meters of her orbit.

On the other hand, no body can fall unless it be attracted, drawn by another body of a more powerful mass.

Beings, animals, objects, adhere to the soil, and weigh upon the Earth, because they are constantly attracted to it by an irresistible force.

Weight and universal attraction are one and the same force.

On the other hand, it can be determined that if an object is left to itself upon the surface of the Earth, it drops 4.90 meters during the first second of its fall.

We also know that attraction diminishes with the square of the distance, and that if we could raise a stone to the height of the Moon, and then abandon it to the attraction of our planet, it would in the first second fall 4.90 meters divided by the square of 60, or 3,600--that is, of 1-1/3 millimeters, exactly the quantity by which the Moon deviates from the straight line she would pursue if the Earth were not influencing her.

The reasoning just stated for the Moon is equally applicable to the Sun.

The distance of the Sun is 23,386 times the radius of the Earth. In order to know how much the intensity of terrestrial weight would be diminished at such a distance, we should look, in the first place, for the square of the number representing the distance--that is, 23,386 multiplied by itself, = 546,905,000. If we divide 4.90 meters, which represents the attractive force of our planet, by this number, we get 9/1000000 of a millimeter, and we see that at the distance of the Sun, the Earth's attraction would really be almost nil.

Now let us do for our planet what we did for its satellite. Let us trace the annual orbit of the terrestrial globe round the central orb, and we shall find that the Earth falls in each second 2.9 millimeters toward the Sun.

This proportion gives the attractive force of the Sun in relation to that of the Earth, and proves that the Sun is 324,000 times more powerful than our world, for 2.9 millimeters divided by 0.000,009 equals 324,000, if worked out into the ultimate fractions neglected here for the sake of simplicity.

A great number of stars have been weighed by the same method.

Their mass is estimated by the movement of a satellite round them, and it is by this method that we are able to affirm that Jupiter is 310 times heavier than the Earth, Saturn 92 times, Neptune 16 times, Uranus 14 times, while Mars is much less heavy, its weight being only two-thirds that of our own.

The planets which have no satellites have been weighed by the perturbations which they cause in other stars, or in the imprudent comets that sometimes tarry in their vicinity. Mercury weighs very much less than the Earth (only 6/100) and Venus about 8/10. So the beautiful star of the evening and morning is not so light as her name might imply, and there is no great difference between her weight and our own.

As the Moon has no secondary body submitted to her influence, her weight has been calculated by reckoning the amount of water she attracts at each tide in the ocean, or by observing the effects of her attraction on the terrestrial globe. When the Moon is before us, in the last quarter, she makes us travel faster, whereas in the first quarter, when she is behind, she delays us.

All the calculations agree in showing us that the orb of night is 81 times less heavy than our planet. There is nearly as much difference in weight between the Earth and the Moon as between an orange and a grape.

* * * * *

Not content with weighing the planets of our system, astronomers have investigated the weight of the stars. How have they been enabled to ascertain the quantity of matter which constitutes these distant Suns--incandescent globes of fire scattered in the depths of space?

They have resorted to the same method, and it is by the study of the attractive influence of a sun upon some other contiguous neighboring star, that the weight of a few of these has been calculated.

Of course this method can only be applied to those double stars of which the distance is known.

It has been discovered that some of the tiny stars that we can hardly see twinkling in the depths of the azure sky are enormous suns, larger and heavier than our own, and millions of times more voluminous than the Earth.

Our planet is only a grain of dust floating in the immensity of Heaven. Yet this atom of infinity is the cradle of an immense creation incessantly renewed, and perpetually transformed by the accumulated centuries.

And what diversity exists in this army of worlds and suns, whose regular harmonious march obeys a mute order....

But we have as yet said nothing about weight on the surface of the worlds, and I see signs of impatience in my readers, for after so much simple if unpoetical demonstration, they will certainly ask me for the explanation that will prove to them that a kilogram transported to Jupiter or Mars would weigh more or less than here.

Give me your attention five minutes longer, and I will restore your faith in the astronomers.

It must not be supposed that objects at the surface of a world like Jupiter, 310 times heavier than our own, weigh 310 times more. That would be a serious error. In that case we should have to assume that a kilogram transported to the surface of the Sun would there weigh 324,000 times more, or 324,000 kilograms. That would be correct if these orbs were of the same dimensions as the Earth. But to speak, for instance, only of the divine Sun, we know that he is 108 times larger than our little planet.

Now, weight at the surface of a celestial body depends not only on its mass, but also on its diameter.

In order to know the weight of any body upon the surface of the Sun, we must argue as follows:

Since a body placed upon the surface of the Sun is 108 times farther from its center than it is upon a globe of the dimensions of the Earth, and since, on the other hand, attraction diminishes with the square of the distance, the intensity of the weight would there be 108 multiplied by 108, or 11,700 times weaker. Now divide the number representing the mass, i.e., 324,000, by this number 11,700, and it results that bodies at the surface of the Sun are 28 times heavier than here. A woman whose weight was 60 kilos would weigh 1,680 kilograms there if organized in the same way as on the Earth, and would find walking very difficult, for at each step she would lift up a shoe that weighed at least ten kilograms.

This reasoning as just stated for the Sun may be applied to the other stars. We know that on the surface of Jupiter the intensity of weight is twice and a third times as great as here, while on Mars it only equals 37/100.

On the surface of Mercury, weight is nearly twice as small again as here. On Neptune it is approximately equal to our own.

With deference to the Selenites, everything is at its lightest on the Moon: a man weighing 70 kilograms on the Earth would not weigh more than 12 kilos there.

So all tastes can be provided for: the only thing to be regretted is that one can not choose one's planet with the same facility as one's residence upon the Earth.

CHAPTER XII

LIFE, UNIVERSAL AND ETERNAL

And now, while thanking my readers for having followed me so far in this descriptive account of the marvels of the Cosmos, I must inquire what philosophical impression has been produced on their minds by these celestial excursions to the other worlds? Are you left indifferent to the pageant of the Heavens? When your imagination was borne away to these distant stars, suns of the infinite, these innumerable stellar systems disseminated through a boundless eternity, did you ask what existed there, what purpose was served by those dazzling spheres, what effects resulted from these forces, radiations, energies? Did you reflect that the elements which upon our little Earth determined a vital activity so prodigious and so varied must needs have spread the waves of an incomparably vaster and more diversified existence throughout the immensities of the Universe? Have you felt that all can not be dead and deserted, as we are tempted by the illusions of our terrestrial senses and of our isolation to believe in the silence of the night: that on the contrary, the real aim of Astronomy, instead of ending with statements of the positions and movements of the stars, is to enable us to penetrate to them, to make us divine, and know, and appreciate their physical constitution, their degree of life and intellectuality in the universal order?

On the Earth, it is Life and Thought that flourish; and it is Life and Thought that we seek again in these starry constellations strewn to Infinitude amid the immeasurable fields of Heaven.

The humble little planet that we inhabit presents itself to us as a brimming cup, overflowing at every outlet. Life is everywhere. From the bottom of the seas, from the valleys to the mountains, from the vegetation that carpets the soil, from the mold in the fields and woods, from the air we breathe, arises an immense, prodigious, and perpetual murmur. Listen! it is the great voice of Nature, the sum of all the unknown and mysterious voices that are forever calling to us, from the ocean waves, from the forest winds, from the 300,000 kinds of insects that are redundant everywhere, and make a lively community on the surface of our globe. A drop of water contains thousands of curious and agile creatures. A grain of dust from the streets of Paris is the home of 130,000 bacteria. If we turn over the soil of a garden, field, or meadow, we find the earthworms working to produce assimilable slime. If we lift a stone in the path, we discover a crawling population. If we gather a flower, detach a leaf, we everywhere find little insects living a parasitic existence. Swarms of midges fly in the sun, the trees of the wood are peopled with nests, the birds sing, and chase each other at play, the lizards dart away at our approach, we trample down the antheaps and the molehills. Life enwraps us in an inexorable encroachment of which we are at once the heroes and the victims, perpetuating itself to its own detriment, as imposed upon it by an eternal reproduction. And this from all time, for the very stones of which we build our houses are full of fossils so prodigiously multiplied that one gram of such stone will often contain millions of shells, marvels of geometrical perfection. The infinitely little is equal to the infinitely great.

Life appears to us as a fatal law, an imperious force which all obey, as the result and the aim of the association of atoms. This is illustrated for us upon the Earth, our only field of direct observation. We must be blind not to see this spectacle, deaf not to hear its reaching. On what pretext could one suppose that our little globe which, as we have seen, has received no privileges from Nature, is the exception; and that the entire Universe, save for one insignificant isle, is devoted to vacancy, solitude, and death?

We have a tendency to imagine that Life can not exist under conditions other than terrestrial, and that the other worlds can only be inhabited on the condition of being similar to our own. But terrestrial nature itself demonstrates to us the error of this way of thinking. We die in the water: fishes die out of the water. Again, short-sighted naturalists affirm categorically that Life is impossible at the bottom of the sea: 1, because it is in complete darkness; 2, because the terrible pressure would burst any organism; 3, because all motion would be impossible there, and so on. Some inquisitive person sends down a dredge, and brings up lovely creatures, so delicate in structure that the daintiest touch must proceed with circumspection. There is no light in these depths: they make it with their own phosphorescence. Other inquirers visit subterranean caverns, and discover animals and plants whose organs have been transformed by adaptation to their gloomy environment.

What right have we to say to the vital energy that radiates round every Sun of the Universe: "Thus far shalt thou come, and no further"? In the name of Science? An absolute mistake. The Known is an infinitesimal island in the midst of the vast ocean of the Unknown. The deep seas which seemed to be a barrier are, as we have seen, peopled with special life. Some one objects: But after all, there is air there, there is oxygen: oxygen is indispensable: a world without oxygen would be a world of death, an eternally sterile desert. Why? Because we have not yet come across beings that can breathe without air, and live without oxygen? Another mistake. Even if we did not know of any, it would not prove that they do not exist. But as it happens, we do know of such: the an 鎙 obia. These beings live without air, without oxygen. Better still: oxygen kills them!

All the evidence goes to show that in interpreting as we ought the spectacle of terrestrial life, and the positive facts acquired by Science, we should enlarge the circle of our conceptions and our judgments, and not limit extra-terrestrial existence to the servile image of what is in existence here below. Terrestrial organic forms are due to local causes upon our planet. The chemical constitution of water and of the atmosphere, temperature, light, density, weight, are so many elements that have gone to form our bodies. Our flesh is composed of carbon, nitrogen, hydrogen, and oxygen combined in the state of water, and of some other elements, among which we may instance sodium chloride (salt). The flesh of animals is not chemically different from our own. All this comes from the water and the air, and returns to them again. The same elements, in very minute quantities, make up all living bodies. The ox that browses on the grass is formed of the same flesh as the man who eats the beef. All organized terrestrial matter is only carbon combined in variable proportions with hydrogen, nitrogen, oxygen, etc.

But we have no right to forbid Nature to act differently in worlds from which carbon is absent. A world, for example, in which silica replaces carbon, silicic acid carbonic acid, might be inhabited by organisms absolutely different from those which exist on the Earth, different not only in form, but also in substance. We already know stars and suns for which spectral analysis reveals a predominance of silica, e.g., Rigel and Deneb. In a world where chlorine predominated, we might expect to find hydrochloric acid, and all the fecund family of chlorides, playing an important part in the phenomena of life. Might not bromine be associated in other formations? Why, indeed, should we draw the line at terrestrial chemistry? What is to prove that these elements are really simple? May not hydrogen, carbon, oxygen, nitrogen, and sulphur all be compounds? Their equivalents are multiples of the first: 1, 6, 8, 14, 16. And is even hydrogen the most simple of the elements? Is not its molecule composed of atoms, and may there not exist a single species of primitive atom, whose geometric arrangement and various associations might constitute the molecules of the so-called simple elements?

In our own solar system we discover the essential differences between certain planets. In the spectrum of Jupiter, for instance, we are aware of the action of an unknown substance that manifests itself by a marked absorption of certain red rays. This gas, which does not exist upon the Earth, is seen still more obviously in the atmospheres of Saturn and Uranus. Indeed, upon this last planet the atmosphere appears, apart from its water vapor, to have no sort of analogy with our own. And in the solar spectrum itself, many of the lines have not yet been identified with terrestrial substances.

The interrelation of the planets is of course incontrovertible, since they are all children of the same parent. But they differ among themselves, not merely in respect of situation, position, volume, mass, density, temperature, atmosphere, but again in physical and chemical constitution. And the point we would now accent is that this diversity should not be regarded as an obstacle to the manifestations of life, but, on the contrary, as a new field open to the infinite fecundity of the universal mother.

When our thoughts take wing, not only to our neighbors, Moon, Venus, Mars, Jupiter, or Saturn, but still more toward the myriads of unknown worlds that gravitate round the suns disseminated in space, we have no plausible reason for imagining that the inhabitants of these other worlds of Heaven resemble us in any way, whether in form, or even in organic substance.

The substance of the terrestrial human body is due to the elements of our planet, and notably to carbon. The terrestrial human form derives from the ancestral animal forms to which it has gradually raised itself by the continuous progress of the transformation of species. To us it seems obvious that we are man or woman, because we have a head, a heart, lungs, two legs, two arms, and so on. Nothing is less a matter of course. That we are constituted as we are, is simply the result of our pro-simian ancestors having also had a head, a heart, lungs, legs, and arms--less elegant than your own, it is true, Madam, but still of the same anatomy. And more and more, by the progress of paleontology, we are delving down to the origin of beings. As certain as it is that the bird derives from the reptile by a process of organic evolution, so certain is it that terrestrial Humanity represents the topmost branches of the huge genealogical tree, whereof all the limbs are brothers, and the roots of which are plunged into the very rudiments of the most elementary and primitive organisms.

The multitude of worlds is surely peopled by every imaginable and unimaginable form. Terrestrial man is endowed with five senses, or perhaps it is better to say six. Why should Nature stop at this point? Why, for instance, may she not have given to certain beings an electrical sense, a magnetic sense, a sense of orientation, an organ able to perceive the ethereal vibrations of the infra-red or ultra-violet, or permitted them to hear at a distance, or to see through walls? We eat and digest like coarse animals, we are slaves to our digestive tube: may there not be worlds in which a nutritive atmosphere enables its fortunate inhabitants to dispense with this absurd process? The least sparrow, even the dusky bat, has an advantage over us in that it can fly through the air. Think how inferior are our conditions, since the man of greatest genius, the most exquisite woman, are nailed to the soil like any vulgar caterpillar before its metamorphosis! Would it be a disadvantage to inhabit a world in which we might fly whither we would; a world of scented luxury, full of animated flowers; a world where the winds would be incapable of exciting a tempest, where several suns of different colors--the diamond glowing with the ruby, or the emerald with the sapphire--would burn night and day (azure nights and scarlet days) in the glory of an eternal spring; with multi-colored moons sleeping in the mirror of the waters, phosphorescent mountains, aerial inhabitants,--men, women, or perhaps of other sexes,--perfect in their forms, gifted with multiple sensibilities, luminous at will, incombustible as asbestos, perhaps immortal, unless they commit suicide out of curiosity? Lilliputian atoms as we are, let us once for all be convinced that our imagination is but sterility, in the midst of an infinitude hardly glimpsed by the telescope.

One important point seems always to be ignored expressly by those who blindly deny the doctrine of the plurality of worlds. It is that this doctrine does not apply more particularly to the present epoch than to any other. Our time is of no importance, no absolute value. Eternity is the field of the Eternal Sower. There is no reason why the other worlds should be inhabited now more than at any other epoch.

What, indeed, is the Present Moment? It is an open trap through which the Future falls incessantly into the gulf of the Past.

The immensity of Heaven bears in its bosom cradles as well as tombs, worlds to come and perished worlds. It abounds in extinct suns, and cemeteries. In all probability Jupiter is not yet inhabited. What does this prove? The Earth was not inhabited during its primordial period: what did that prove to the inhabitants of Mars or of the Moon, who were perhaps observing it at that epoch, a few million years ago?

To pretend that our globe must be the only inhabited world because the others do not resemble it, is to reason, not like a philosopher, but, as we remarked before, like a fish. Every rational fish ought to assume that it is impossible to live out of water, since its outlook and its philosophy do not extend beyond its daily life. There is no answer to this order of reasoning, except to advise a little wider perception, and extension of the too narrow horizon of habitual ideas.

For us the resources of Nature may be considered infinite, and "positive" science, founded upon our senses only, is altogether inadequate, although it is the only possible basis of our reasoning. We must learn to see with the eyes of our spirit.

As to the planetary systems other than our own, we are no longer reduced to hypotheses. We already know with certainty that our Sun is no exception, as was suggested, and is still maintained, by some theorists. The discovery in itself is curious enough.

It is surely an exceptional situation that, given a sidereal system composed of a central sun, and of one or more stars gravitating round him, the plane of such a system should fall just within our line of vision, and that it should revolve in such a way that the globes of which it is composed pass exactly between this sun and ourselves in turning round him, eclipsing him more or less during this transit. As, on the other hand, the eclipses would be our only means of determining the existence of these unknown planets (save indeed from perturbation, as in the case of Sirius and Procyon), it might have seemed quixotic to hope for like conditions in order to discover solar systems other than our own. But these exceptional circumstances have reproduced themselves at different parts of the Heavens.

Thus, for instance, we have seen that the variable star Algol owes its variations in brilliancy, which reduce it from second to fourth magnitude every sixty-nine hours, to the interposition of a body between itself and the Earth, and celestial mechanics has already been able to determine accurately the orbit of this body, its dimensions and its mass, and even the flattening of the sun Algol. Here, then, is a system in which we know the sun and an enormous planet, whose revolution is effected in sixty-nine hours with extreme rapidity, as measured by the spectroscope.

The star [delta] of Cepheus is in the same case: it is an orb eclipsed in a period of 129 hours, and its eclipsing planet also revolves in the plane of our vision. The variable star in Ophiuchus has an analogous system, and observation has already revealed a great number of others.

Since, then, a certain number of solar systems differing from our own have been revealed, as it were in section, to terrestrial observation, this affords us sufficient evidence of the existence of an innumerable quantity of solar systems scattered through the immensities of space, and we are no longer reduced to conjecture.

On the other hand, analysis of the motions of several stars, such as Sirius, Procyon, Alta 飏, proves that these distant orbs have companions,--planets not yet discovered by the telescope, and that perhaps never will be discovered, because they are obscure, and lost in the radiation of the star.

* * * * *

Some savants have asserted that Life can not germinate if the conditions of the environment differ too much from terrestrial conditions.

This hypothesis is purely gratuitous, and we will now discuss it.

In order to examine what is happening on the Earth, let us mount the ladder of time for a moment, to follow the evolutions of Nature.

There was an epoch when the Earth did not exist. Our planet, the future world of our habitation, slept in the bosom of the solar nebula.

At last it came to birth, this cherished Earth, a gaseous, luminous ball, poor reflection of the King of Orbs, its parent. Millions of years rolled by before the condensation and cooling of this new globe were sufficiently transformed to permit life to manifest itself in its most rudimentary aspects.

The first organic forms of the protoplasm, the first aggregations of cells, the protozoons, the zoophytes or plant-animals, the gelatinous mussels of the still warm seas, were succeeded by the fishes, then by the reptiles, the birds, the mammals, and lastly man, who at present occupies the top of the genealogical tree, and crowns the animal kingdom.

Humanity is comparatively young upon the Earth. We may attribute some thousands of centuries of existence to it ... and some five years of reason!

The terrestrial organisms, from the lowest up to man, are the resultant of the forces in action at the surface of our planet. The earliest seem to have been produced by the combinations of carbon with hydrogen and nitrogen; they were, so to speak, without animation, save for some very rudimentary sensibility; the sponges, corals, polyps, and medus? give us a notion of these primitive beings. They were formed in the tepid waters of the primary epoch. As long as there were no continents, no islands emerging from the level of the universal ocean, there were no beings breathing in the air. The first aquatic creatures were succeeded by the amphibia, the reptiles. Later on were developed the mammals and the birds.

What, again, do we not owe to the plant-world of the primary epoch, of the secondary epoch, of the tertiary epoch, which slowly prepared the good nutritious soil of to-day, in which the roses flourish, and the peach and strawberry ripen?

Before it gave birth to a Helen or a Cleopatra, life manifested itself under the roughest forms, and in the most varied conditions. A long-period comet passing in sight of the Earth from time to time would have seen modifications of existence in each of its transits, in accordance with a slow evolution, corresponding to the variation of the conditions of existence, and progressing incessantly, for if Life is the goal of nature, Progress is the supreme law.

The history of our planet is the history of life, with all its metamorphoses. It is the same for all the worlds, with some exceptions of orbs arrested in their development.

The constitution of living beings is in absolute relation with the substances of which they are composed, the environment in which they move, temperature, light, weight, density, the length of day and night, the seasons, etc.--in a word, with all the cosmographic elements of a world.

If, for example, we compare between themselves two worlds such as the Earth and Neptune, utterly different from the point of view of distance from the Sun, we could not for an instant suppose that organic structures could have followed a parallel development on these planets. The average temperature must be much lower on Neptune than on the Earth, and the same holds for intensity of light. The years and seasons there are 165 times longer than with us, the density of matter is three times as weak, and weight is, on the contrary, a little greater. Under conditions so different from our own, the activities of Nature would have to translate themselves under other forms. And doubtless the elementary bodies would not be found there in the same proportions. Consequently we have to conclude that organs and senses would not be the same there as here. The optic nerve, for instance, which has formed and developed here from the rudimentary organ of the trilobite to the marvels of the human eye, must be incomparably more sensitive upon Neptune than in our dazzling solar luminosity, in order to perceive radiations that we do not perceive here. In all probability, it is replaced there by some other organ. The lungs, functioning there in another atmosphere, are different from our own. So, too, for the stomach and digestive organs. Corporeal forms, animal and human, can not resemble those which exist upon the Earth.

Certain savants contend that if the conditions differed too much from terrestrial conditions, life could not be produced there at all. Yet we have no right to limit the powers of Nature to the narrow bounds of our sphere of observation, and to pretend that our planet and our Humanity are the type of all the worlds. That is a hypothesis as ridiculous as it is childish.

Do not let us be "personal," like children, and old people who never see beyond their room. Let us learn to live in the Infinite and the Eternal.

From this larger point of view, the doctrine of the plurality of worlds is the complement and the natural crown of Astronomy. What interests us most in the study of the Universe is surely to know what goes on there.

* * * * *

These considerations show that, in all the ages, what really constitutes a planet is not its skeleton but the life that vibrates upon its surface.

And again, if we analyze things, we see that for the Procession of Nature, life is all, and matter nothing.

What has become of our ancestors, the millions of human beings who preceded us upon this globe? Where are their bodies? What is left of them? Search everywhere. Nothing is left but the molecules of air, water, dust, atoms of hydrogen, nitrogen, oxygen, carbon, etc., which are incorporated in turn in the organism of every living being.

The whole Earth is a vast cemetery, and its finest cities are rooted in the catacombs. But now, in crossing Paris, I passed for at least the thousandth time near the Church of St. Germain-l'Auxerrois, and was obliged to turn out of the direct way, on account of excavations. I looked down, and saw that immediately below the pavement, they had just uncovered some stone coffins still containing the skeletons that had reposed there for ten centuries. From time immemorial the passers-by had trampled them unwittingly under foot. And I reflected that it is much the same in every quarter of Paris. Only yesterday, some Roman tombs and a coin with the effigy of Nero were found in a garden near the Observatory.

And from the most general standpoint of Life, the whole world is in the same case, and even more so, seeing that all that exists, all that lives, is formed of elements that have already been incorporated in other beings, no longer living. The roses that adorn the bosom of the fair ... but I will not enlarge upon this topic.

And you, so strong and virile, of what elements is your splendid body formed? Where have the elements you absorb to-day in respiration and assimilation been drawn from, what lugubrious adventures have they been subject to? Think away from it: do not insist on this point: on no account consider it....

And yet, let us dwell on it, since this reality is the most evident demonstration of the ideal; since what exists is you, is all of us, is Life; and matter is only its substance, like the materials of a house, and even less so, since its particles only pass rapidly through the framework of our bodies. A heap of stones does not make a house. Quintillions of tons of materials would not represent the Earth or any other world.

Yes, what really exists, what constitutes a complete orb, is the city of Life. Let us recognize that the flower of life flourishes on the surface of our planet, embellishing it with its perfume; that it is just this life that we see and admire,--of which we form part,--and which is the raison d'être of things; that matter floats, and crosses, and crosses back again, in the web of living beings,--and the reality, the goal, is not matter--it is the life matter is employed upon.

Yes, matter passes, and being also, after sharing in the concerted symphony of life.

And indeed everything passes rapidly!

What irrepressible grief, what deep melancholy, what ineffaceable regrets we feel, when as age comes on we look back, when we see our friends fallen upon the road one after the other, above all when we visit the beloved scenes of our childhood, those homes of other years, that witnessed our first start in terrestrial existence, our first games, our first affections--those affections of childhood that seemed eternal--when we wander over those fields and valleys and hills, when we see again the landscape whose aspect has hardly changed, and whose image is so intimately linked with our first impressions. There near this fireside the grandfather danced us on his knee, and told us blood-curdling stories; here the kind grandmother came to see if we were comfortably tucked in, and not likely to fall out of the big bed; in this little wood, along these alleys that seemed endless, we spread our nets for birds; in this stream we fished for crayfish; there on the path we played at soldiers with our elders, who were always captains; on these slopes we found rare stones and fossils, and mysterious petrifactions; on this hill we admired the fine sunsets, the appearance of the stars, the form of the constellations. There we began to live, to think, to love, to form attachments, to dream, to question every problem, to breathe intellectually and physically. And now, where is this beloved grandfather? the good grandmother? where are all whom we knew in infancy? where are our dreams of childhood? Winged thoughts still seem to flutter in the air, and that is all. People, caresses, voices, all have gone and vanished. The cemetery has closed over them all. There is a silent void. Were all those fine and sunny hours an illusion? Was it only to weep one day over this negation that our childish hearts were so tenderly attached to these fleeting shadows? Is there nothing, down the long length of human history, but eternal delusion?

It is here, above all, that we find ourselves in presence of the greatest problems. Life is the goal, it is Life that produces the conditions of Thought. Without Thought, where would be the Universe?

We feel that without life and thought, the Universe would be an empty theater, and Astronomy itself, sublime science, a vain research. We feel that this is the truth, veiled as yet to actual science, and that human races kindred with our own exist there in the immensities of space. Yes, we feel that this is truth.

But we would fain go a little further in our knowledge of the universe, and penetrate in some measure the secret of our destinies. We would know if these distant and unknown Humanities are not attached to us by mysterious cords, if our life, which will assuredly be extinguished at some definite moment here below, will not be prolonged into the regions of Eternity.

A moment ago we said that nothing is left of the body. Millions of organisms have lived, there are no remains of them. Air, water, smoke, dust. Memento, homo, quia pulvis es et in pulverem revertebis. Remember oh man! that dust thou art, and unto dust thou shalt return, says the priest to the faithful, when he scatters the ashes on the day after the carnival.

The body disappears entirely. It goes where the corpse of Caesar went an hour after the extinction of his pyre. Nor will there be more remains of any of us. And the whole of Humanity, and the Earth itself, will also disappear one day. Let no one talk of the Progress of Humanity as an end! That would be too gross a decoy.

If the soul were also to disappear in smoke, what would be left of the vital and intellectual organization of the world? Nothing.

On this hypothesis, all would be reduced to nothing.

Our reason is not immense, our terrestrial faculties are sufficiently limited, but this reason and these faculties suffice none the less to make us feel the improbability, the absurdity, of this hypothesis, and we reject it as incompatible with the sublime grandeur of the spectacle of the universe.

Undoubtedly, Creation does not seem to concern itself with us. It proceeds on its inexorable course without consulting our sensations. With the poet we regret the implacable serenity of Nature, opposing the irony of its smiling splendor to our mourning, our revolts, and our despair.

How brief a time suffices for all things to change! Serene-fronted Nature, too soon you will forget!... in your metamorphoses ruthlessly snapping the cords that bind our hearts together!

Others will pass where we pass; we have arrived, and others will arrive after us: the thought sketched out by our souls will be pursued by theirs ... and they will not find the solution of it.

For no one here begins or finishes: the worst are as the best of humans; we all awake at the same moment of the dream: we all begin in this world, and end otherwhere.

Reply, sweet valley, reply, solitude; O Nature, sheltering in this splendid desert, when we are both asleep, and cast by the tomb into the attitude of pensive death.

Will you to the last verge be so insensible, that, knowing us lost, and dead with our loves, you will pursue your cheerful feast, and smile, and sing always?

Yes, mortals may say that when they are sleeping in the grave, spring and summer will still smile and sing; husband and wife may ask themselves if they will meet again some day, in another sphere; but do we not feel that our destinies can not be terminated here, and that short of absolute and final nonentity for everything, they must be renewed beyond, in that starry Heaven to which every dream has flown instinctively since the first origins of Humanity?

As our planet is only a province of the Infinite Heavens, so our actual existence is only a stage in Eternal Life. Astronomy, by giving us wings, conducts us to the sanctuary of truth. The specter of death has departed from our Heaven. The beams of every star shed a ray of hope into our hearts. On each sphere Nature chants the paen of Life Eternal.

THE END